Further Pure Mathematics 1

Edexcel AS and A-level Modular Mathematics

Greg Attwood
Lee Cope
Bronwen Moran
Laurence Pateman
Keith Pledger
Geoff Staley
Dave Wilkins

Contents

About this book

This book is designed to provide you with the best preparation possible for your Edexcel FP1 unit examination:

- This is Edexcel's own course for the GCE specification.
- Written by senior examiners.
- The LiveText CD-ROM in the back of the book contains even more resources to support you through the unit.
- A matching FP1 revision guide is also available.

Finding your way around the book

Brief chapter overview and 'links' to underline the importance of mathematics: to the real world, to your study of further units and to your career

Every few chapters, a review exercise helps you consolidate your learning

Detailed contents list shows which parts of the FP1 specification are covered in each section

Contents

ii

2

After completing this chapter you should be able to
- find approximations to the solutions of equations of the form $f(x) = 0$ using:
  - internal bisection
  - linear interpolation
  - the Newton–Raphson process.

Numerical solutions of equations

Numerical methods are used in science and engineering to help solve problems. These problems are normally modelled using computers.
The numerical methods used lead to approximate solutions to the many equations that need to be solved.
Weather forecasters use numerical methods to predict the weather, both in the immediate future (a few hours) and up to a few weeks ahead.

32

1 Review Exercise

1 $z_1 = 2 + i$, $z_2 = 3 + 4i$. Find the modulus and the tangent of the argument of each of
a $z_1 z_2^*$ **b** $\frac{z_1}{z_2}$ E

2 **a** Show that the complex number $\frac{2+3i}{5+i}$ can be expressed in the form $\lambda(1 + i)$, stating the value of λ.
b Hence show that $\left(\frac{2+3i}{5+i}\right)^4$ is real and determine its value. E

3 $z_1 = 5 + i$, $z_2 = -2 + 3i$
a Show that $|z_1|^2 = 2|z_2|^2$.
b Find $\arg(z_1 z_2)$. E

4 **a** Find, in the form $p + iq$ where p and q are real, the complex number z which satisfies the equation $\frac{3z-1}{2-i} = \frac{4}{1+2i}$.
b Show on a single Argand diagram the points which represent z and z^*.
c Express z and z^* in modulus–argument form, giving the arguments to the nearest degree. E

5 $z_1 = -1 + i\sqrt{3}$, $z_2 = \sqrt{3} + i$
a Find **i** $\arg z_1$ **ii** $\arg z_2$.
b Express $\frac{z_1}{z_2}$ in the form $a + ib$, where a and b are real, and hence find $\arg\left(\frac{z_1}{z_2}\right)$.
c Verify that $\arg\left(\frac{z_1}{z_2}\right) = \arg z_1 - \arg z_2$. E

6 **a** Find the two square roots of $3 - 4i$ in the form $a + ib$, where a and b are real.
b Show the points representing the two square roots of $3 - 4i$ in a single Argand diagram. E

7 The complex number z is $-9 + 17i$.
a Show z on an Argand diagram.
b Calculate $\arg z$, giving your answer in radians to two decimal places.
c Find the complex number w for which $zw = 25 + 35i$, giving your answer in the form $p + iq$, where p and q are real. E

63

Each section begins with a statement of what is covered in the section

Concise learning points

Step-by-step worked examples – they are model solutions and include examiners hints

Past examination questions are marked 'E'

Chapter 5

5.1 You can use the $\sum$ notation.

- The sigma notation is a very useful, and concise way, to define a series. It makes further study of series more manageable.
- As you saw in Book C1, Chapter 6, where the $\sum$ notation was first introduced,
  $\sum_{r=1}^{n} U_r = U_1 + U_1 + U_2 + U_3 + \ldots U_n$, where U_r is a function of r.
- It is also used to mean 'the sum of the series'. For example, $\sum_{r=1}^{n} r^2 = 1^2 + 2^2 + 3^2 + \ldots + n^2$, but you will also see that $\sum_{r=1}^{n} r^2 = \frac{n}{6}(n+1)(2n+1)$, the sum of the series (this result will be proved in Chapter 6).
- Series such as arithmetic series, geometric series and the binomial series, which you have already studied, can all be written in $\sum$ notation: for example, $\sum_{k=0}^{n-1} ar^k$ 'sums up' the geometric series $a + ar + ar^2 + \ldots + ar^{n-1}$.

Example 1

Write out the series defined by the following

a $\sum_{r=1}^{n}(2r-1)$

b $\sum_{r=0}^{n-1}(2r+1)$

c $\sum_{k=1}^{n}(k^2+2)$

These expressions represent a general term in the series.

a $\sum_{r=1}^{n}(2r-1) = (2 \times 1 - 1) + (2 \times 2 - 1) + (2 \times 3 - 1) + (2 \times 4 - 1) + \ldots + (2n-1) = 1 + 3 + 5 + 7 + \ldots + (2n-1)$

b $\sum_{r=0}^{n-1}(2r+1) = (2 \times 0 + 1) + (2 \times 1 + 1) + (2 \times 2 + 1) + (2 \times 3 + 1) + \ldots + [2(n-1) + 1] = 1 + 3 + 5 + 7 + \ldots + (2n-1)$

The same series can be expressed in different ways in $\sum$ notation.

c $\sum_{k=1}^{n}(k^2+2) = (1^2+2) + (2^2+2) + (3^2+2) + (4^2+2) + \ldots + (n^2+2) = 3 + 6 + 11 + 18 + \ldots + (n^2+2)$

108

Example 2

Write these series using the $\sum$ notation.

a $3 + 5 + 7 + 9 + 11 + 13$

b $2 + 5 + 10 + \ldots + (n^2 + 1)$

c $1 \times 2 + 2 \times 3 + 3 \times 4 + \ldots + (n-2)(n-1)$

a $3 + 5 + 7 + 9 + 11 + 13$ is the sum of the odd numbers from 3 to 13.
An odd number may be represented by $(2r - 1)$, where r is an integer.
The values of r corresponding to 3 and 13 are 2 and 7 respectively, so the series can be written as $\sum_{r=2}^{7}(2r-1)$.

Equally $\sum_{r=1}^{6}(2r+1)$ could be used.

b $2 + 5 + 10 + \ldots + (n^2 + 1) = \sum_{r=1}^{n}(r^2+1)$

c The general term of the series $1 \times 2 + 2 \times 3 + 3 \times 4 + \ldots + (n-2)(n-1)$ can be written as $r(r+1)$. Using this expression the first term corresponds to $r = 1$ and the final term corresponds to $r = (n-2)$, so we can write $\sum_{r=1}^{n-2} r(r+1)$.

Exercise 5A

1 Write out each of the following as a sum of terms, and hence calculate the sum of the series.

a $\sum_{r=1}^{10} r$ **b** $\sum_{p=3}^{8} p^2$ **c** $\sum_{r=1}^{10} r^3$

d $\sum_{p=1}^{10}(2p^2+3)$ **e** $\sum_{r=0}^{5}(7r+1)^2$ **f** $\sum_{i=1}^{4} 2i(3-4i^2)$

2 Write each of the following as a sum of terms, showing the first three terms and the last term.

a $\sum_{r=1}^{n}(7r-1)$ **b** $\sum_{r=1}^{n}(2r^3+1)$

c $\sum_{j=1}^{n}(j-4)(j+4)$ **d** $\sum_{p=3}^{k} p(p+3)$

109

Each section ends with an exercise – the questions are carefully graded so they increase in difficulty and gradually bring you up to standard

Each chapter has a different colour scheme, to help you find the right chapter quickly

Each chapter ends with a mixed exercise and a summary of key points.

At the end of the book there is an examination-style paper.

LiveText software

The LiveText software gives you additional resources: Solutionbank and Exam café. Simply turn the pages of the electronic book to the page you need, and explore!

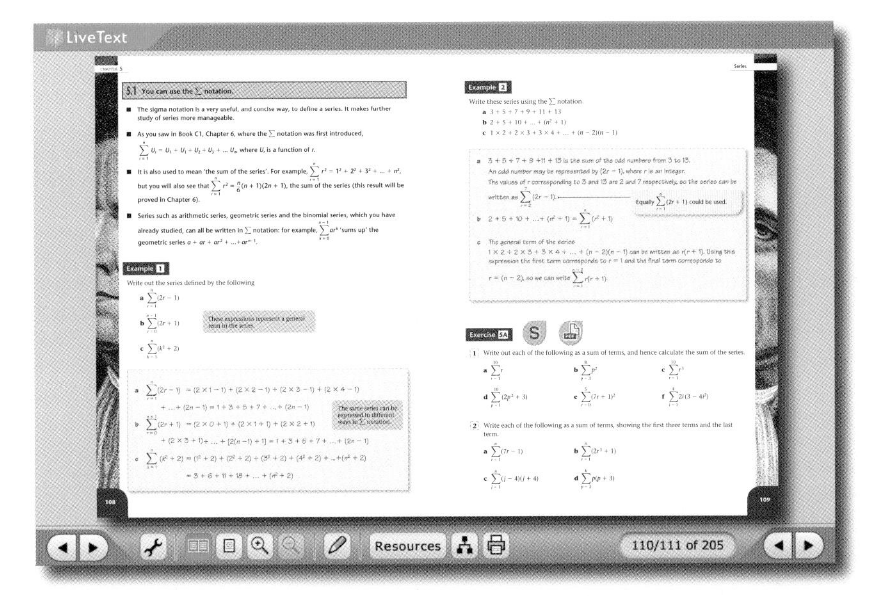

Unique Exam café feature:

- Relax and prepare – revision planner; hints and tips; common mistakes
- Refresh your memory – revision checklist; language of the examination; glossary
- Get the result! – fully worked examination-style paper with chief examiner's commentary

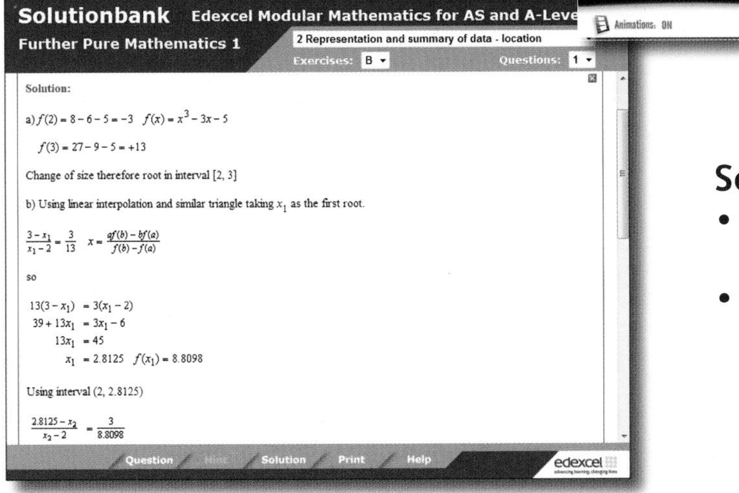

Solutionbank

- Hints and solutions to every question in the textbook
- Solutions and commentary for all review exercises and the practice examination paper

Pearson Education Limited, a company incorporated in England and Wales, having its registered office at Edinburgh Gate, Harlow, Essex, CM20 2JE. Registered company number: 872828

15 13 14 13
12 11 10 9 8

British Library Cataloguing in Publication Data is available from the British Library on request.

ISBN 978 0 435519 230

Edited by Susan Gardner
Typeset by Tech-Set Ltd, Gateshead
Illustrated by Tech-Set Ltd, Gateshead
Cover design by Christopher Howson
Picture research by Chrissie Martin
Index by Indexing Specialists (UK)
Cover photo/illustration © Science Photo Library / Laguna Design
Printed in China (CTPS/08)

Acknowledgements
The author and publisher would like to thank the following individuals and organisations for permission to reproduce photographs:

Fotolia / Solar Wind Studios p**1**; Digital Vision p**32**; Alamy / David Cheshire p**41**; Moviestore Collection p**72**; Science Photo Library / Sheila Terry p**107**; Corbis / Hoge Noorden / epa p**122**

Every effort has been made to contact copyright holders of material reproduced in this book. Any omissions will be rectified in subsequent printings if notice is given to the publishers.

After completing this chapter you should be able to:

- add, subtract, multiply and divide **complex numbers**
- find the **modulus** and **argument** of a complex number
- show complex numbers on an **Argand diagram**
- solve equations that have **complex roots**.

Complex numbers

Although complex numbers may seem to have few direct links with real-world quantities, there are areas of application in which the idea of a complex number is extremely useful. For example, the strength of an electromagnetic field, which has both an electric and a magnetic component, can be described by using a complex number. Other areas in which the mathematics of complex numbers is a valuable tool include signal processing, fluid dynamics and quantum mechanics.

The *Aurora Borealis* (Northern Lights) are part of the Earth's electromagnetic field.

1.1 You can use real and imaginary numbers.

When solving a quadratic equation in Unit C1, you saw how the **discriminant** of the equation could be used to find out about the type of roots.

For the equation $ax^2 + bx + c = 0$, the discriminant is $b^2 - 4ac$.
If $b^2 - 4ac > 0$, there are two different real roots.
If $b^2 - 4ac = 0$, there are two equal real roots.
If $b^2 - 4ac < 0$, there are no real roots.

In the case $b^2 - 4ac < 0$, the problem is that you reach a situation where you need to find the square root of a negative number, which is not 'real'.

To solve this problem, another type of number called an 'imaginary number' is used.

The 'imaginary number' $\sqrt{(-1)}$ is called i (or sometimes j in electrical engineering), and sums of real and imaginary numbers, such as 3 + 2i, are known as **complex numbers**.

- **A complex number is written in the form $a + bi$.**
- **You can add and subtract complex numbers.**
- $\sqrt{(-1)} = \text{i}$
- **An imaginary number is a number of the form bi, where b is a real number ($b \in \mathbb{R}$).**

Example 1

Write $\sqrt{(-36)}$ in terms of i.

$\sqrt{(-36)} = \sqrt{(36 \times -1)} = \sqrt{36}\sqrt{(-1)} = 6\text{i}$

Example 2

Write $\sqrt{(-28)}$ in terms of i.

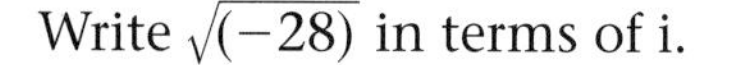

$\sqrt{(-28)} = \sqrt{(28 \times -1)} = \sqrt{28}\sqrt{(-1)} = \sqrt{4}\sqrt{7}\sqrt{(-1)} = 2\sqrt{7}\text{i}$ or $2\text{i}\sqrt{7}$ or $(2\sqrt{7})\text{i}$

This can be written as $2\text{i}\sqrt{7}$ or $(2\sqrt{7})\text{i}$ to avoid confusion with $2\sqrt{7\text{i}}$.

Example 3

Solve the equation $x^2 + 9 = 0$.

$x^2 = -9$

$x = \pm\sqrt{(-9)} = \pm\sqrt{(9 \times -1)} = \pm\sqrt{9}\sqrt{(-1)} = \pm 3\text{i}$

$x = \pm 3\text{i} \qquad (x = +3\text{i}, x = -3\text{i})$

Note that just as $x^2 = 9$ has two roots +3 and −3, $x^2 = -9$ also has two roots +3i and −3i.

- A complex number is a number of the form $a + bi$, where $a \in \mathbb{R}$ and $b \in \mathbb{R}$.
- For the complex number $a + bi$, a is called the **real part** and b is called the **imaginary part.**
- The complete set of complex numbers is called $\mathbb{C}$.

Example 4

Solve the equation $x^2 + 6x + 25 = 0$.

Method 1 (Completing the square)

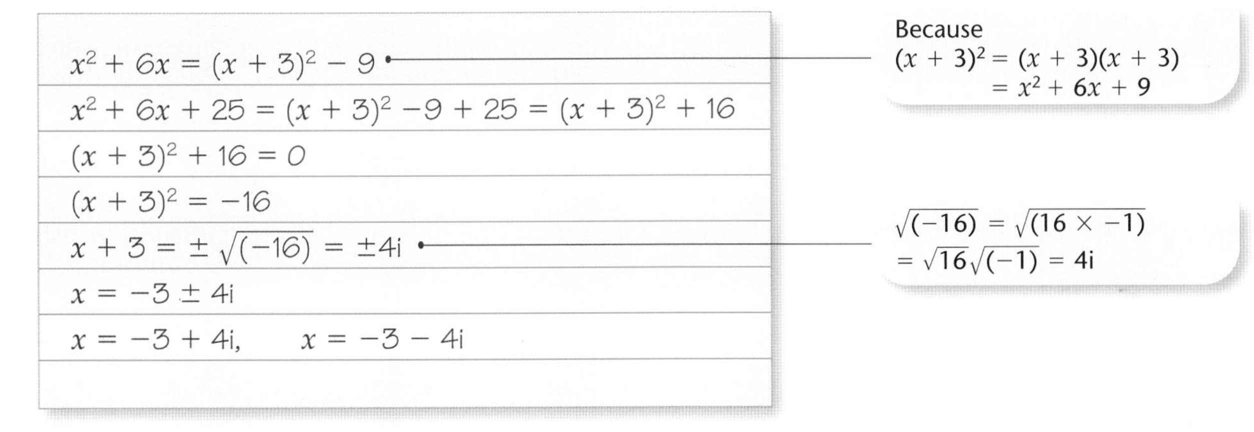

$x^2 + 6x = (x + 3)^2 - 9$

$x^2 + 6x + 25 = (x + 3)^2 - 9 + 25 = (x + 3)^2 + 16$

$(x + 3)^2 + 16 = 0$

$(x + 3)^2 = -16$

$x + 3 = \pm\sqrt{(-16)} = \pm 4i$

$x = -3 \pm 4i$

$x = -3 + 4i, \quad x = -3 - 4i$

Because $(x + 3)^2 = (x + 3)(x + 3) = x^2 + 6x + 9$

$\sqrt{(-16)} = \sqrt{(16 \times -1)} = \sqrt{16}\sqrt{(-1)} = 4i$

Method 2 (Quadratic formula)

$$x = \frac{-6 \pm \sqrt{(6^2 - 4 \times 1 \times 25)}}{2} = \frac{-6 \pm \sqrt{(-64)}}{2}$$

$\sqrt{(-64)} = \pm 8i$

$$x = \frac{-6 \pm 8i}{2} = -3 \pm 4i$$

$x = -3 + 4i, \quad x = -3 - 4i$

Using $x = \dfrac{-b \pm \sqrt{(b^2 - 4ac)}}{2a}$

$\sqrt{(-64)} = \sqrt{(64 \times -1)} = \sqrt{64}\sqrt{(-1)} = 8i$

- In a complex number, the real part and the imaginary part cannot be combined to form a single term.
- You can add complex numbers by adding the real parts and adding the imaginary parts.
- You can subtract complex numbers by subtracting the real parts and subtracting the imaginary parts.

Example 5

Simplify, giving your answer in the form $a + bi$, where $a \in \mathbb{R}$ and $b \in \mathbb{R}$.

a $(2 + 5i) + (7 + 3i)$ **b** $(3 - 4i) + (-5 + 6i)$ **c** $2(5 - 8i)$

d $(1 + 8i) - (6 + i)$ **e** $(2 - 5i) - (5 - 11i)$ **f** $(2 + 3i) - (2 - 3i)$

a $(2 + 5i) + (7 + 3i) = (2 + 7) + i(5 + 3) = 9 + 8i$

Add real parts and add imaginary parts.

b $(3 - 4i) + (-5 + 6i) = (3 - 5) + i(-4 + 6) = -2 + 2i$

c $2(5 - 8i) = 10 - 16i$

This is the same as $(5 - 8i) + (5 - 8i)$

d $(1 + 8i) - (6 + i) = (1 - 6) + i(8 - 1) = -5 + 7i$

Subtract real parts and subtract imaginary parts.

e $(2 - 5i) - (5 - 11i) = (2 - 5) + i(-5 - (-11)) = -3 + 6i$

f $(2 + 3i) - (2 - 3i) = (2 - 2) + i(3 - (-3)) = 6i$

The answer has no real part. This is called purely imaginary.

Exercise 1A

Simplify, giving your answer in the form $a + bi$, where $a \in \mathbb{R}$ and $b \in \mathbb{R}$.

1 $(5 + 2i) + (8 + 9i)$

2 $(4 + 10i) + (1 - 8i)$

3 $(7 + 6i) + (-3 - 5i)$

4 $(2 - i) + (11 + 2i)$

5 $(3 - 7i) + (-6 + 7i)$

6 $(20 + 12i) - (11 + 3i)$

7 $(9 + 6i) - (8 + 10i)$

8 $(2 - i) - (-5 + 3i)$

9 $(-4 - 6i) - (-8 - 8i)$

10 $(-1 + 5i) - (-1 + i)$

11 $(3 + 4i) + (4 + 5i) + (5 + 6i)$

12 $(-2 - 7i) + (1 + 3i) - (-12 + i)$

13 $(18 + 5i) - (15 - 2i) - (3 + 7i)$

14 $2(7 + 2i)$

15 $3(8 - 4i)$

16 $7(1 - 3i)$

17 $2(3 + i) + 3(2 + i)$

18 $5(4 + 3i) - 4(-1 + 2i)$

19 $\left(\frac{1}{2} + \frac{1}{3}i\right) + \left(\frac{5}{2} + \frac{5}{3}i\right)$

20 $(3\sqrt{2} + i) - (\sqrt{2} - i)$

Write in the form bi, where $b \in \mathbb{R}$.

21 $\sqrt{(-9)}$

22 $\sqrt{(-49)}$

23 $\sqrt{(-121)}$

24 $\sqrt{(-10\,000)}$

25 $\sqrt{(-225)}$

26 $\sqrt{(-5)}$

27 $\sqrt{(-12)}$

28 $\sqrt{(-45)}$

29 $\sqrt{(-200)}$

30 $\sqrt{(-147)}$

Solve these equations.

31 $x^2 + 2x + 5 = 0$

32 $x^2 - 2x + 10 = 0$

33 $x^2 + 4x + 29 = 0$

34 $x^2 + 10x + 26 = 0$

35 $x^2 - 6x + 18 = 0$

36 $x^2 + 4x + 7 = 0$

37 $x^2 - 6x + 11 = 0$

38 $x^2 - 2x + 25 = 0$

39 $x^2 + 5x + 25 = 0$

40 $x^2 + 3x + 5 = 0$

1.2 You can multiply complex numbers and simplify powers of i.

- **You can multiply complex numbers using the same technique as you use for multiplying brackets in algebra, and you can simplify powers of i.**
- **Since $i = \sqrt{(-1)}$, $i^2 = -1$**

Example 6

Multiply (2 + 3i) by (4 + 5i)

$$(2 + 3i)(4 + 5i) = 2(4 + 5i) + 3i(4 + 5i)$$

Multiply the two brackets as you would in algebra.

$$= 8 + 10i + 12i + 15i^2$$

$$= 8 + 10i + 12i - 15$$

Use the fact that $i^2 = -1$.

$$= (8 - 15) + (10i + 12i)$$

Add real parts and add imaginary parts.

$$= -7 + 22i$$

Example 7

Express $(7 - 4i)^2$ in the form $a + bi$.

$$(7 - 4i)(7 - 4i) = 7(7 - 4i) - 4i(7 - 4i)$$

Multiply the two brackets as you would in algebra.

$$= 49 - 28i - 28i + 16i^2$$

$$= 49 - 28i - 28i - 16$$

Use the fact that $i^2 = -1$.

$$= (49 - 16) + (-28i - 28i)$$

Add real parts and add imaginary parts.

$$= 33 - 56i$$

Example 8

Simplify $(2 - 3i)(4 - 5i)(1 + 3i)$

$(2 - 3i)(4 - 5i) = 2(4 - 5i) - 3i(4 - 5i)$

$= 8 - 10i - 12i + 15i^2 = 8 - 10i - 12i - 15 = -7 - 22i$

$(-7 - 22i)(1 + 3i) = -7(1 + 3i) - 22i(1 + 3i)$

$= -7 - 21i - 22i - 66i^2 = 59 - 43i$

First multiply two of the brackets.

Then multiply the result by the third bracket.

Example 9

Simplify

a i^3 **b** i^4 **c** $(2i)^5$

a $i^3 = i \times i \times i = i^2 \times i = -i$

$i^2 = -1$

b $i^4 = i \times i \times i \times i = i^2 \times i^2 = -1 \times -1 = 1$

c $(2i)^5 = 2i \times 2i \times 2i \times 2i \times 2i = 32(i \times i \times i \times i \times i)$

$= 32(i^2 \times i^2 \times i) = 32 \times -1 \times -1 \times i = 32i$

First multiply the 2s (2^5).

Exercise 1B

Simplify these, giving your answer in the form $a + bi$.

1 $(5 + i)(3 + 4i)$

2 $(6 + 3i)(7 + 2i)$

3 $(5 - 2i)(1 + 5i)$

4 $(13 - 3i)(2 - 8i)$

5 $(-3 - i)(4 + 7i)$

6 $(8 + 5i)^2$

7 $(2 - 9i)^2$

8 $(1 + i)(2 + i)(3 + i)$

9 $(3 - 2i)(5 + i)(4 - 2i)$

10 $(2 + 3i)^3$

Simplify.

11 i^6

12 $(3i)^4$

13 $i^5 + i$

14 $(4i)^3 - 4i^3$

15 $(1 + i)^8$

Hint: Use the binomial theorem.

1.3 You can find the complex conjugate of a complex number.

- **You can write down the complex conjugate of a complex number, and you can divide two complex numbers by using the complex conjugate of the denominator.**
- **The complex number $a - bi$ is called the complex conjugate of the complex number $a + bi$.**
- **The complex numbers $a + bi$ and $a - bi$ are called a complex conjugate pair.**
- **The complex conjugate of z is called z^*, so if $z = a + bi$, $z^* = a - bi$.**

Example 10

Write down the complex conjugate of

a $2 + 3i$ **b** $5 - 2i$ **c** $\sqrt{3} + i$ **d** $1 - i\sqrt{5}$

a $2 - 3i$ **b** $5 + 2i$

c $\sqrt{3} - i$ **d** $1 + i\sqrt{5}$

Just change the sign of the imaginary part (from + to –, or – to +).

Example 11

Find $z + z^*$ and zz^*, given that

a $z = 3 + 5i$ **b** $z = 2 - 7i$ **c** $z = 2\sqrt{2} + i\sqrt{2}$

a
$z^* = 3 - 5i$

$z + z^* = (3 + 5i) + (3 - 5i) = (3 + 3) + i(5 - 5) = 6$ — Note that $z + z^*$ is real.

$zz^* = (3 + 5i)(3 - 5i) = 3(3 - 5i) + 5i(3 - 5i)$

$= 9 - 15i + 15i - 25i^2 = 9 + 25 = 34$ — Note that zz^* is real.

b
$z^* = 2 + 7i$

$z + z^* = (2 - 7i) + (2 + 7i) = (2 + 2) + i(-7 + 7) = 4$ — Note that $z + z^*$ is real.

$zz^* = (2 - 7i)(2 + 7i) = 2(2 + 7i) - 7i(2 + 7i)$

$= 4 + 14i - 14i - 49i^2 = 4 + 49 = 53$ — Note that zz^* is real.

c
$z^* = 2\sqrt{2} - i\sqrt{2}$

$z + z^* = (2\sqrt{2} + i\sqrt{2}) + (2\sqrt{2} - i\sqrt{2})$

$= (2\sqrt{2} + 2\sqrt{2}) + i(\sqrt{2} - \sqrt{2}) = 4\sqrt{2}$ — Note that $z + z^*$ is real.

$zz^* = (2\sqrt{2} + i\sqrt{2})(2\sqrt{2} - i\sqrt{2})$

$= 2\sqrt{2}(2\sqrt{2} - i\sqrt{2}) + i\sqrt{2}(2\sqrt{2} - i\sqrt{2})$

$= 8 - 4i + 4i - 2i^2 = 8 + 2 = 10$ — Note that zz^* is real.

Example 12

Simplify $(10 + 5i) \div (1 + 2i)$

$$(10 + 5i) \div (1 + 2i) = \frac{10 + 5i}{1 + 2i} \times \frac{1 - 2i}{1 - 2i}$$

The complex conjugate of the denominator is $1 - 2i$. Multiply numerator and denominator by this.

$$\frac{10 + 5i}{1 + 2i} \times \frac{1 - 2i}{1 - 2i} = \frac{(10 + 5i)(1 - 2i)}{(1 + 2i)(1 - 2i)}$$

$$(10 + 5i)(1 - 2i) = 10(1 - 2i) + 5i(1 - 2i)$$
$$= 10 - 20i + 5i - 10i^2$$
$$= 20 - 15i$$

$$(1 + 2i)(1 - 2i) = 1(1 - 2i) + 2i(1 - 2i)$$
$$= 1 - 2i + 2i - 4i^2 = 5$$

$$(10 + 5i) \div (1 + 2i) = \frac{20 - 15i}{5} = 4 - 3i$$

Divide each term in the numerator by 5.

Example 13

Simplify $(5 + 4i) \div (2 - 3i)$

$$(5 + 4i) \div (2 - 3i) = \frac{5 + 4i}{2 - 3i} \times \frac{2 + 3i}{2 + 3i}$$

The complex conjugate of the denominator is $2 + 3i$. Multiply numerator and denominator by this.

$$\frac{5 + 4i}{2 - 3i} \times \frac{2 + 3i}{2 + 3i} = \frac{(5 + 4i)(2 + 3i)}{(2 - 3i)(2 + 3i)}$$

$$(5 + 4i)(2 + 3i) = 5(2 + 3i) + 4i(2 + 3i)$$
$$= 10 + 15i + 8i + 12i^2$$
$$= -2 + 23i$$

$$(2 - 3i)(2 + 3i) = 2(2 + 3i) - 3i(2 + 3i)$$
$$= 4 + 6i - 6i - 9i^2 = 13$$

$$(5 + 4i) \div (2 - 3i) = \frac{-2 + 23i}{13} = -\frac{2}{13} + \frac{23}{13}i$$

Divide each term in the numerator by 13.

The division process shown in Examples 12 and 13 is similar to the process used to divide surds. (See C1 Section 1.8.)
For surds the denominator is rationalised. For complex numbers the denominator is made real.

- **If the roots α and β of a quadratic equation are complex, α and β will always be a complex conjugate pair.**

- **If the roots of the equation are α and β, the equation is $(x - \alpha)(x - \beta) = 0$**
$(x - \alpha)(x - \beta) = x^2 - \alpha x - \beta x + \alpha\beta = x^2 - (\alpha + \beta)x + \alpha\beta$

Example 14

Find the quadratic equation that has roots 3 + 5i and 3 − 5i.

For this equation $\alpha + \beta = (3 + 5i) + (3 - 5i) = 6$

and $\alpha\beta = (3 + 5i)(3 - 5i) = 9 + 15i - 15i - 25i^2 = 34$

The equation is $x^2 - 6x + 34 = 0$

This is a useful method to remember, although it is not required knowledge for the FP1 exam.

Exercise 1C

1 Write down the complex conjugate z^* for

a $z = 8 + 2i$ **b** $z = 6 - 5i$

c $z = \frac{2}{3} - \frac{1}{2}i$ **d** $z = \sqrt{5} + i\sqrt{10}$

2 Find $z + z^*$ and zz^* for

a $z = 6 - 3i$ **b** $z = 10 + 5i$

c $z = \frac{3}{4} + \frac{1}{4}i$ **d** $z = \sqrt{5} - 3i\sqrt{5}$

Find these in the form $a + bi$.

3 $(25 - 10i) \div (1 - 2i)$

4 $(6 + i) \div (3 + 4i)$

5 $(11 + 4i) \div (3 + i)$

6 $\dfrac{1 + i}{2 + i}$

7 $\dfrac{3 - 5i}{1 + 3i}$

8 $\dfrac{3 + 5i}{6 - 8i}$

9 $\dfrac{28 - 3i}{1 - i}$

10 $\dfrac{2 + i}{1 + 4i}$

11 $\dfrac{(3 - 4i)^2}{1 + i}$

Given that $z_1 = 1 + i$, $z_2 = 2 + i$ and $z_3 = 3 + i$, find answers for questions 12–14 in the form $a + bi$.

12 $\dfrac{z_1 z_2}{z_3}$

13 $\dfrac{(z_2)^2}{z_1}$

14 $\dfrac{2z_1 + 5z_3}{z_2}$

15 Given that $\dfrac{5 + 2i}{z} = 2 - i$, find z in the form $a + bi$.

16 Simplify $\dfrac{6 + 8i}{1 + i} + \dfrac{6 + 8i}{1 - i}$, giving your answer in the form $a + bi$.

17 The roots of the quadratic equation $x^2 + 2x + 26 = 0$ are α and β.
Find **a** α and β **b** $\alpha + \beta$ **c** $\alpha\beta$

18 The roots of the quadratic equation $x^2 - 8x + 25 = 0$ are α and β.
Find **a** α and β **b** $\alpha + \beta$ **c** $\alpha\beta$

19 Find the quadratic equation that has roots $2 + 3i$ and $2 - 3i$.

20 Find the quadratic equation that has roots $-5 + 4i$ and $-5 - 4i$.

1.4 You can represent complex numbers on an Argand diagram.

- You can represent **complex numbers** on a diagram, called an **Argand diagram.**
- A **real number** can be represented as a point on a **straight line (a number line, which has one dimension).**
- A **complex number**, having two components (real and imaginary), can be represented as a point in a **plane (two dimensions).**
- The complex number $z = x + iy$ is represented by the point (x, y), where x and y are **Cartesian coordinates.**
- The Cartesian coordinate diagram used to represent complex numbers is called an Argand diagram.
- The x-axis in the Argand Diagram is called the **real axis** and the y-axis is called the **imaginary axis.**

Example 15

The complex numbers $z_1 = 2 + 5i$, $z_2 = 3 - 4i$ and $z_3 = -4 + i$ are represented by the points A, B and C respectively on an Argand diagram. Sketch the Argand diagram.

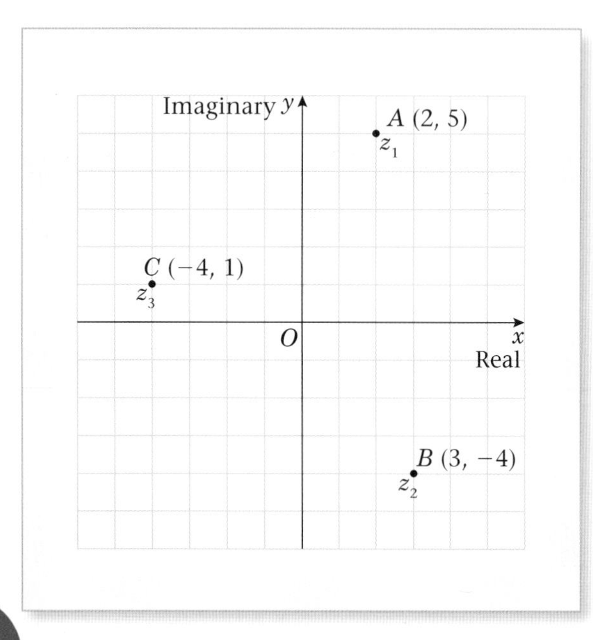

For $z_1 = 2 + 5i$, plot $(2, 5)$.
For $z_2 = 3 - 4i$, plot $(3, -4)$.
For $z_3 = -4 + i$, plot $(-4, 1)$.

Example 16

Show the complex conjugates $z_1 = 4 + 2\text{i}$ and $z_2 = 4 - 2\text{i}$ on an Argand diagram.

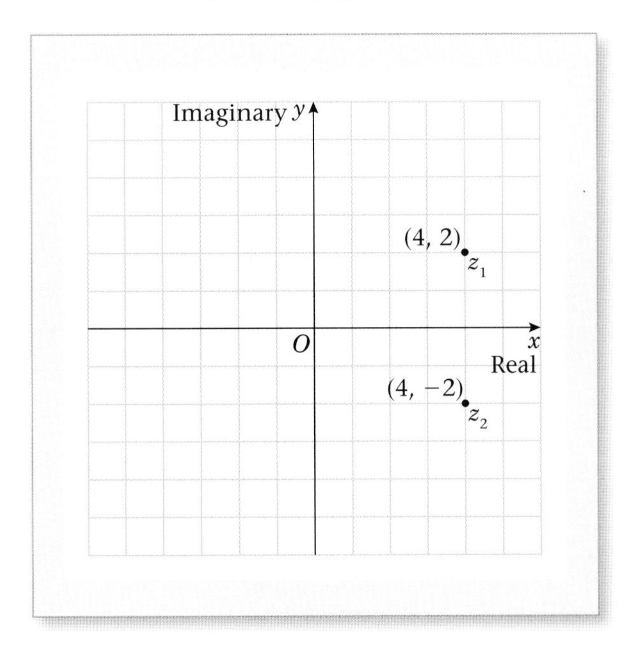

Note that complex conjugates will always be placed symmetrically above and below the real axis.

The complex number $z = x + \text{i}y$ can also be represented by the vector $\overrightarrow{OP}$, where O is the origin and P is the point (x, y) on the Argand diagram.

Example 17

Show the complex numbers $z_1 = 2 + 5\text{i}$, $z_2 = 3 - 4\text{i}$ and $z_3 = -4 + \text{i}$ on an Argand diagram.

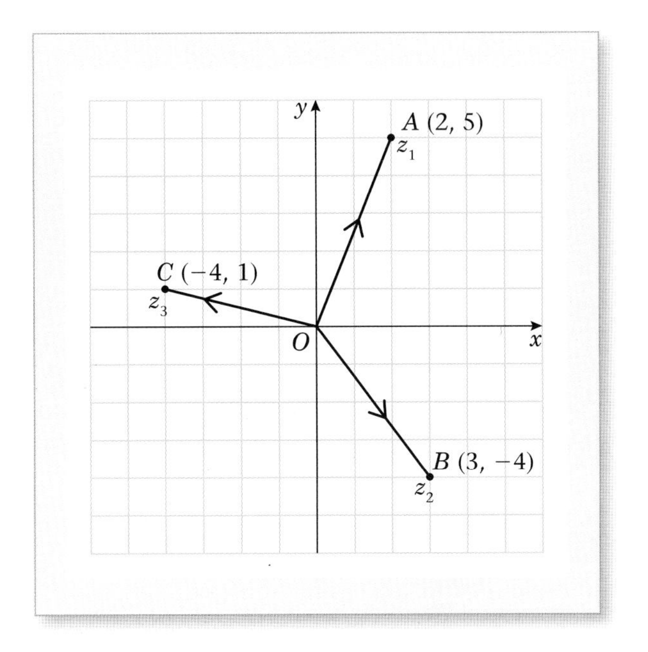

For $z_1 = 2 + 5\text{i}$, show the vector from (0, 0) to (2, 5).
Similarly for z_2 and z_3.

If you label the diagram with letters A, B and C, make sure that you show which letter represents which vector.

Example 18

The complex numbers $z_1 = 7 + 24\text{i}$ and $z_2 = -2 + 2\text{i}$ are represented by the vectors $\overrightarrow{OA}$ and $\overrightarrow{OB}$ respectively on an Argand diagram (where O is the origin). Draw the diagram and calculate $|\overrightarrow{OA}|$ and $|\overrightarrow{OB}|$.

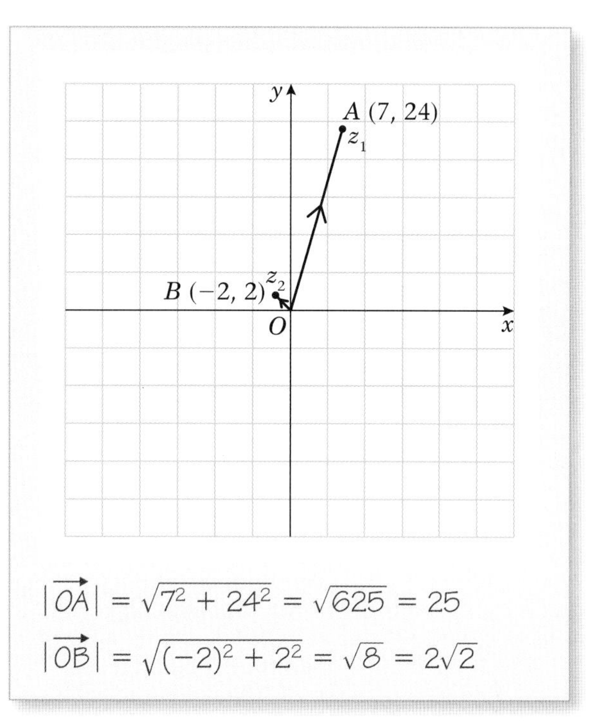

$|\overrightarrow{OA}| = \sqrt{7^2 + 24^2} = \sqrt{625} = 25$

$|\overrightarrow{OB}| = \sqrt{(-2)^2 + 2^2} = \sqrt{8} = 2\sqrt{2}$

■ **Addition of complex numbers can be represented on the Argand diagram by the addition of their respective vectors on the diagram.**

Example 19

$z_1 = 4 + \text{i}$ and $z_2 = 3 + 3\text{i}$. Show z_1, z_2 and $z_1 + z_2$ on an Argand diagram.

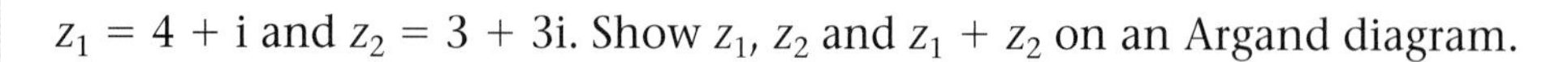

$z_1 + z_2 = (4 + 3) + \text{i}(1 + 3) = 7 + 4\text{i}$

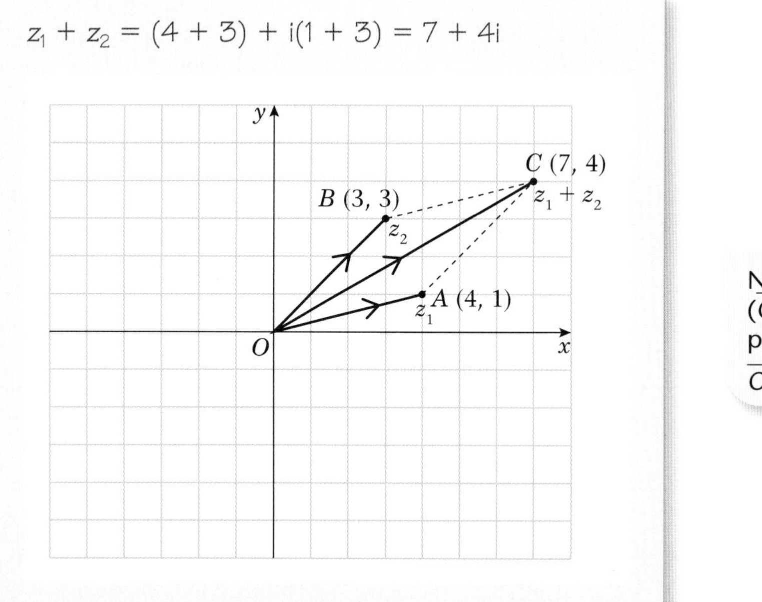

Note that the vector for $z_1 + z_2$ ($\overrightarrow{OC}$) is the diagonal of the parallelogram. This is because $\overrightarrow{OC} = \overrightarrow{OA} + \overrightarrow{AC} = \overrightarrow{OA} + \overrightarrow{OB}$.

Example 20

$z_1 = 6 - 2\text{i}$ and $z_2 = -1 + 4\text{i}$. Show z_1, z_2 and $z_1 + z_2$ on an Argand diagram.

$z_1 + z_2 = (6 - 1) + \text{i}(-2 + 4) = 5 + 2\text{i}$

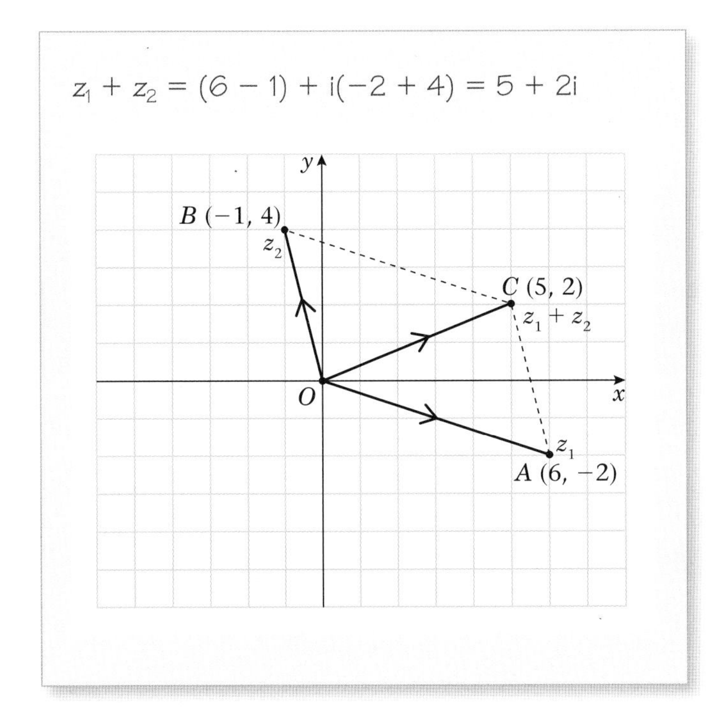

Note that the vector for $z_1 + z_2$ ($\overrightarrow{OC}$) is the diagonal of the parallelogram. This is because $\overrightarrow{OC} = \overrightarrow{OA} + \overrightarrow{AC} = \overrightarrow{OA} + \overrightarrow{OB}$.

Example 21

$z_1 = 2 + 5\text{i}$ and $z_2 = 4 + 2\text{i}$. Show z_1, z_2 and $z_1 - z_2$ on an Argand diagram.

$z_1 - z_2 = (2 - 4) + \text{i}(5 - 2) = -2 + 3\text{i}$

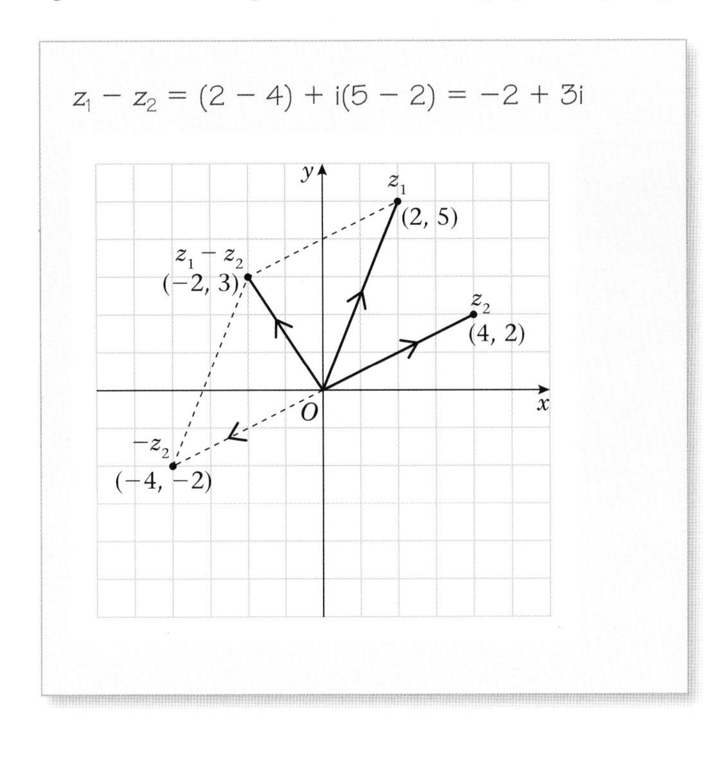

$z_1 - z_2 = z_1 + (-z_2)$.
The vector for $-z_2$ is shown by the dotted line on the diagram.

Exercise 1D

1 Show these numbers on an Argand diagram.

a $7 + 2i$ **b** $5 - 4i$

c $-6 - i$ **d** $-2 + 5i$

e $3i$ **f** $\sqrt{2} + 2i$

g $-\frac{1}{2} + \frac{5}{2}i$ **h** -4

2 Given that $z_1 = -1 - i$, $z_2 = -5 + 10i$ and $z_3 = 3 - 4i$,

a find z_1z_2, z_1z_3 and $\frac{z_2}{z_3}$ in the form $a + ib$.

b show z_1, z_2, z_3, z_1z_2, z_1z_3 and $\frac{z_2}{z_3}$ on an Argand diagram.

3 Show the roots of the equation $x^2 - 6x + 10 = 0$ on an Argand diagram.

4 The complex numbers $z_1 = 5 + 12i$, $z_2 = 6 + 10i$, $z_3 = -4 + 2i$ and $z_4 = -3 - i$ are represented by the vectors $\overrightarrow{OA}$, $\overrightarrow{OB}$, $\overrightarrow{OC}$ and $\overrightarrow{OD}$ respectively on an Argand diagram. Draw the diagram and calculate $|\overrightarrow{OA}|$, $|\overrightarrow{OB}|$, $|\overrightarrow{OC}|$ and $|\overrightarrow{OD}|$.

5 $z_1 = 11 + 2i$ and $z_2 = 2 + 4i$. Show z_1, z_2 and $z_1 + z_2$ on an Argand diagram.

6 $z_1 = -3 + 6i$ and $z_2 = 8 - i$. Show z_1, z_2 and $z_1 + z_2$ on an Argand diagram.

7 $z_1 = 8 + 4i$ and $z_2 = 6 + 7i$. Show z_1, z_2 and $z_1 - z_2$ on an Argand diagram.

8 $z_1 = -6 - 5i$ and $z_2 = -4 + 4i$. Show z_1, z_2 and $z_1 - z_2$ on an Argand diagram.

1.5 You can find the value of *r*, the modulus of a complex number *z*, and the value of *θ*, the argument of *z*.

- **Consider the complex number 3 + 4i, represented on an Argand diagram by the point *A*, or by the vector $\overrightarrow{OA}$.**

The length *OA* or $|\overrightarrow{OA}|$, the magnitude of vector $|\overrightarrow{OA}|$, is found by Pythagoras' theorem:

$|\overrightarrow{OA}| = \sqrt{3^2 + 4^2} = \sqrt{25} = 5$

This number is called the modulus of the complex number 3 + 4i.

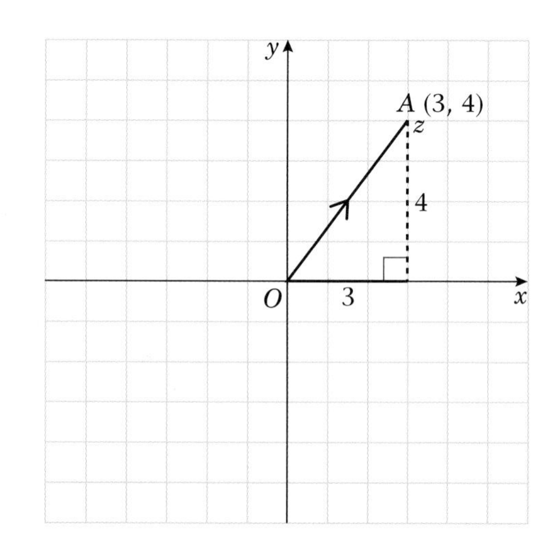

The **modulus** of the complex number $z = x + iy$ is given by $\sqrt{x^2 + y^2}$.

The modulus of the complex number $z = x + iy$ is written as r or $|z|$ or $|x + iy|$, so $r = \sqrt{x^2 + y^2}$.

$|z| = \sqrt{x^2 + y^2}$.

$|x + iy| = \sqrt{x^2 + y^2}$.

■ The modulus of any non-zero complex number is positive.

Consider again the complex number $z = 3 + 4i$.

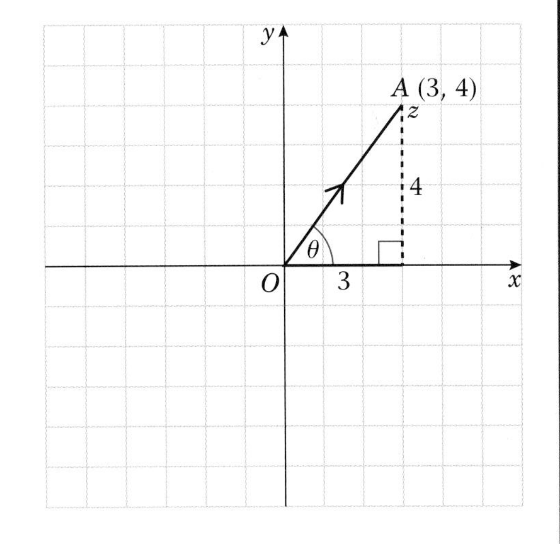

By convention, angles are measured from the positive x-axis (or the positive real axis), anticlockwise being positive.

The angle θ shown on the Argand diagram, measured from the positive real axis, is found by trigonometry:

$\tan \theta = \frac{4}{3}$,

$\theta = \arctan \frac{4}{3} \approx 0.927$ radians

This angle is called the argument of the complex number $3 + 4i$.

The **argument** of the complex number $z = x + iy$ is the angle θ between the positive real axis and the vector representing $\mathbf{z}$ on the Argand diagram.

For the argument θ of the complex number $z = x + iy$, $\tan \theta = \frac{y}{x}$.

The argument θ of any complex number is such that $-\pi < \theta \leqslant \pi$ (or $-180° < \theta \leqslant 180°$). (This is sometimes referred to as the **principal argument**).

The argument of a complex number z is written as **arg z**.

The argument θ of a complex number is usually given in radians.

It is important to remember that the position of the complex number on the Argand diagram (the quadrant in which it appears) will determine whether its argument is positive or negative and whether its argument is acute or obtuse.

The following examples illustrate this.

Example 22

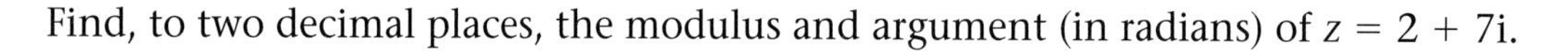

Find, to two decimal places, the modulus and argument (in radians) of $z = 2 + 7\text{i}$.

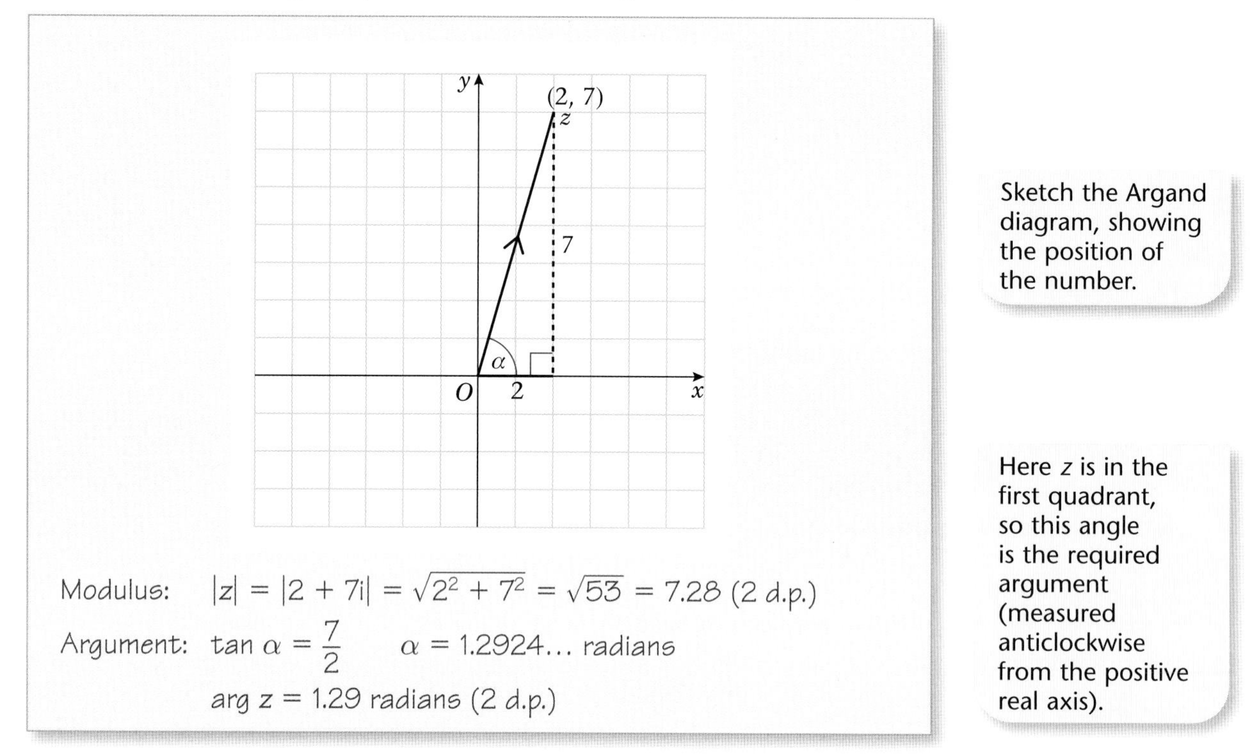

Modulus: $|z| = |2 + 7\text{i}| = \sqrt{2^2 + 7^2} = \sqrt{53} = 7.28$ (2 d.p.)

Argument: $\tan \alpha = \dfrac{7}{2}$ $\quad \alpha = 1.2924\ldots$ radians

$\arg z = 1.29$ radians (2 d.p.)

Sketch the Argand diagram, showing the position of the number.

Here z is in the first quadrant, so this angle is the required argument (measured anticlockwise from the positive real axis).

Example 23

Find, to two decimal places, the modulus and argument (in radians) of $z = -5 + 2\text{i}$.

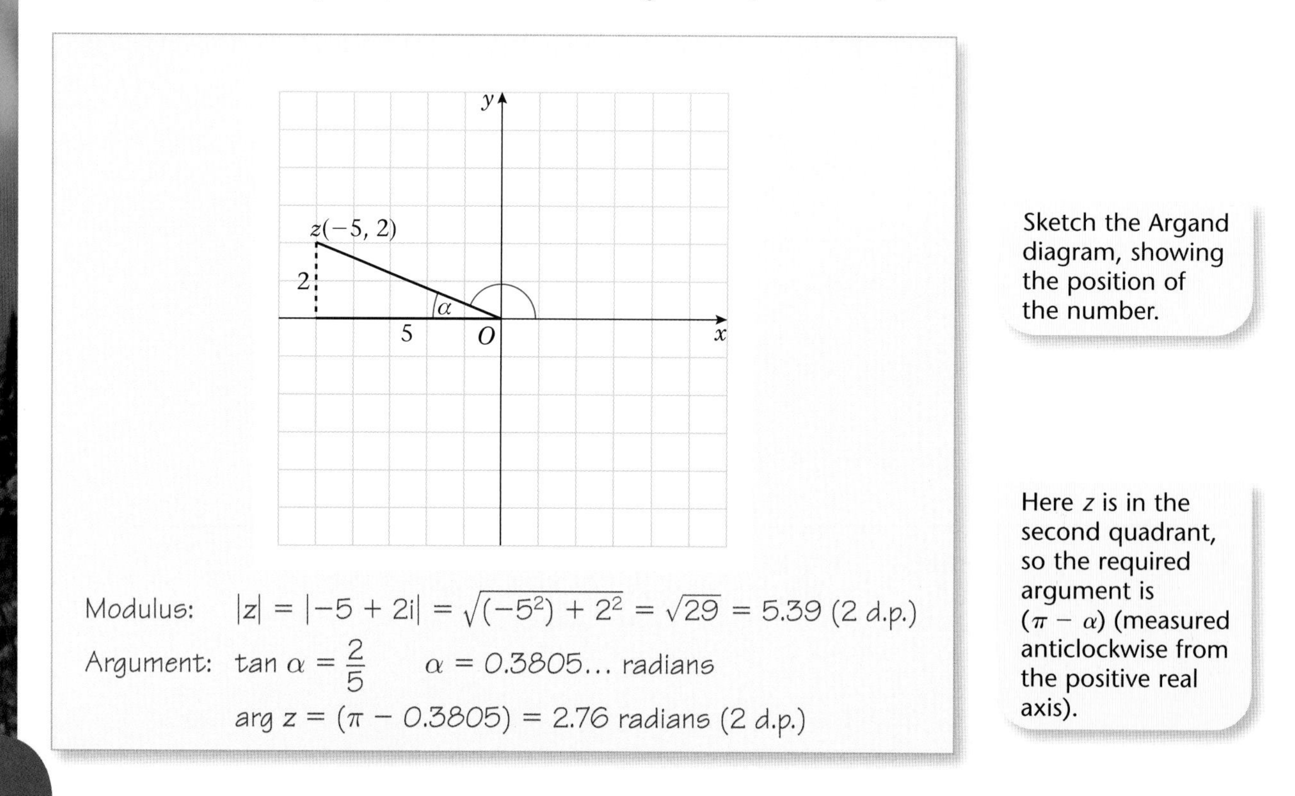

Modulus: $|z| = |-5 + 2\text{i}| = \sqrt{(-5^2) + 2^2} = \sqrt{29} = 5.39$ (2 d.p.)

Argument: $\tan \alpha = \dfrac{2}{5}$ $\quad \alpha = 0.3805\ldots$ radians

$\arg z = (\pi - 0.3805) = 2.76$ radians (2 d.p.)

Sketch the Argand diagram, showing the position of the number.

Here z is in the second quadrant, so the required argument is $(\pi - \alpha)$ (measured anticlockwise from the positive real axis).

Example 24

Find, to two decimal places, the modulus and argument (in radians) of $z = -4 - \text{i}$.

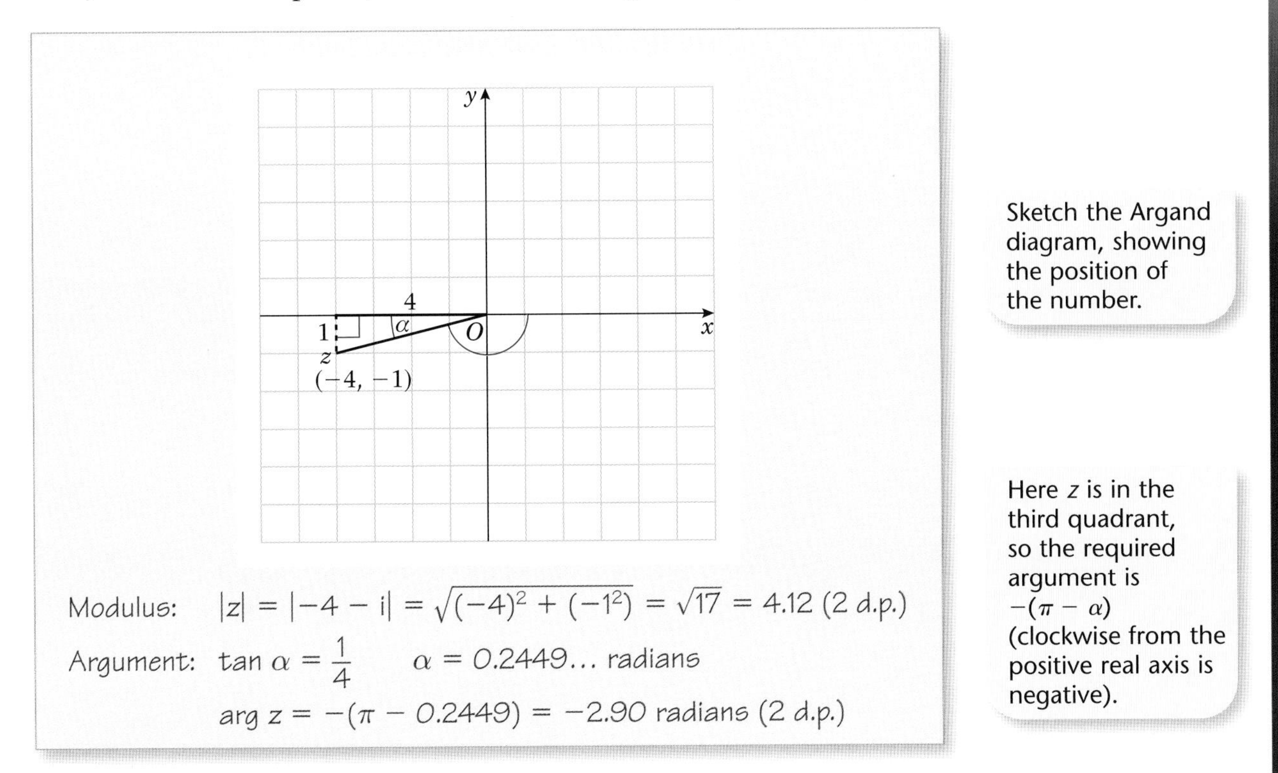

Example 25

Find, to two decimal places, the modulus and argument (in radians) of $z = 3 - 7\text{i}$.

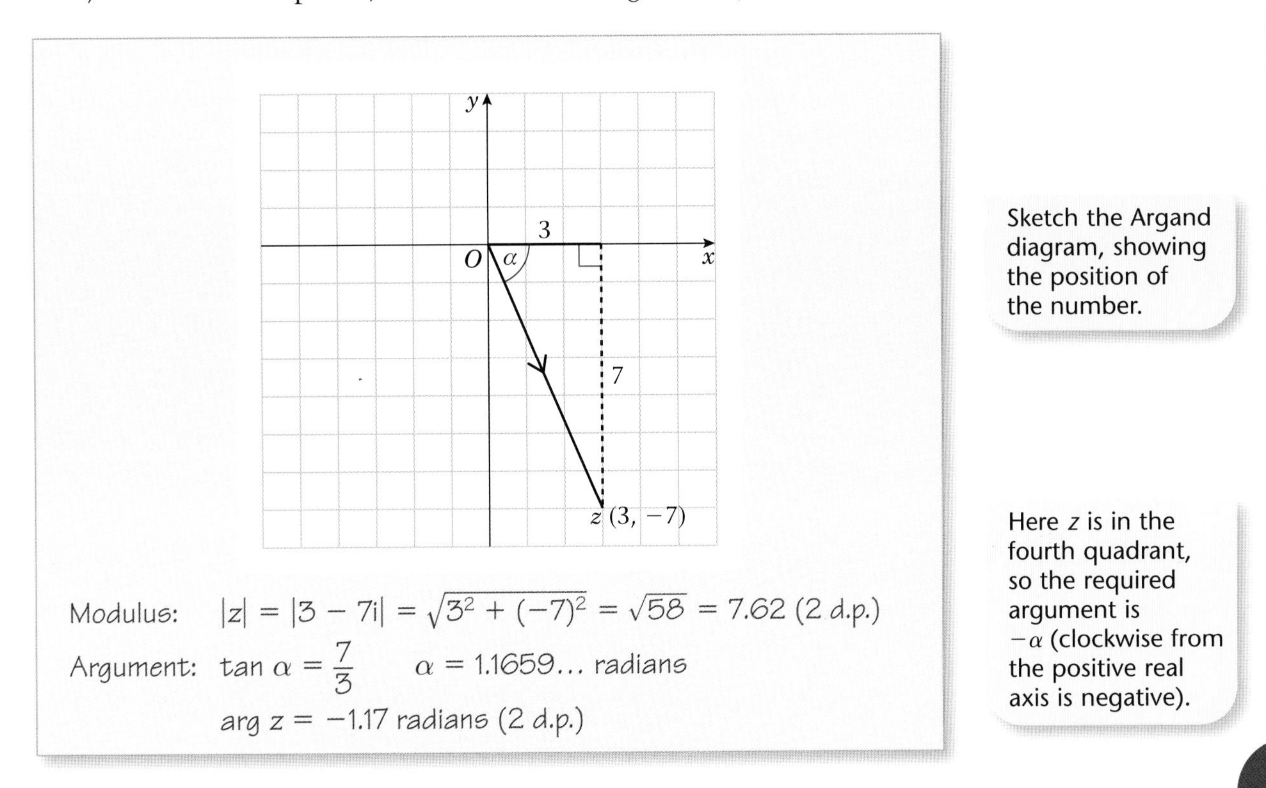

Example 26

Find the exact values of the modulus and argument (in radians) of $z = -1 + \text{i}$.

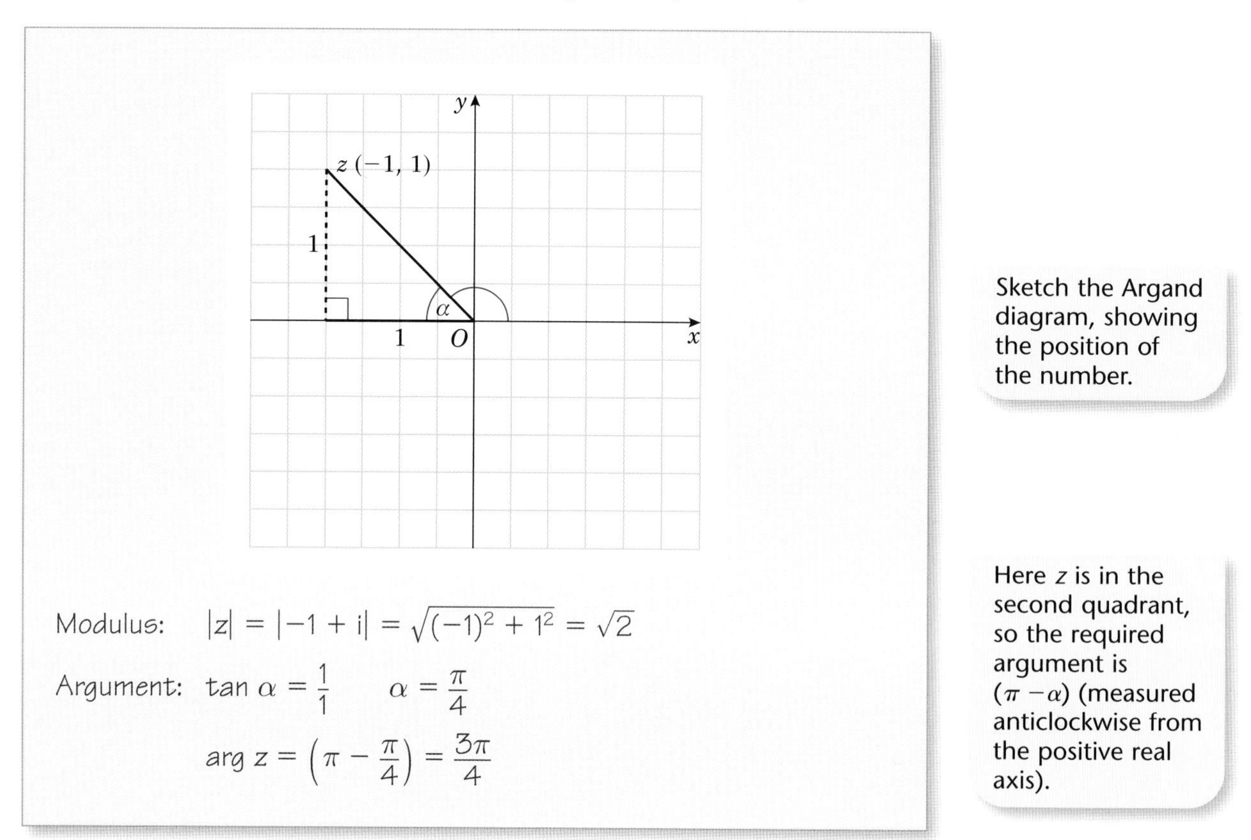

Sketch the Argand diagram, showing the position of the number.

Modulus: $|z| = |-1 + \text{i}| = \sqrt{(-1)^2 + 1^2} = \sqrt{2}$

Argument: $\tan \alpha = \frac{1}{1} \qquad \alpha = \frac{\pi}{4}$

$\arg z = \left(\pi - \frac{\pi}{4}\right) = \frac{3\pi}{4}$

Here z is in the second quadrant, so the required argument is $(\pi - \alpha)$ (measured anticlockwise from the positive real axis).

Exercise 1E

Find the modulus and argument of each of the following complex numbers, giving your answers exactly where possible, and to two decimal places otherwise.

1 $12 + 5\text{i}$

2 $\sqrt{3} + \text{i}$

3 $-3 + 6\text{i}$

4 $2 - 2\text{i}$

5 $-8 - 7\text{i}$

6 $-4 + 11\text{i}$

7 $2\sqrt{3} - \text{i}\sqrt{3}$

8 $-8 - 15\text{i}$

1.6 You can find the modulus–argument form of the complex number z.

- **The modulus–argument form of the complex number $z = x + iy$ is $z = r(\cos\theta + i\sin\theta)$ where r is a positive real number and θ is an angle such that $-\pi < \theta \leqslant \pi$ (or $-180° < \theta \leqslant 180°$)**

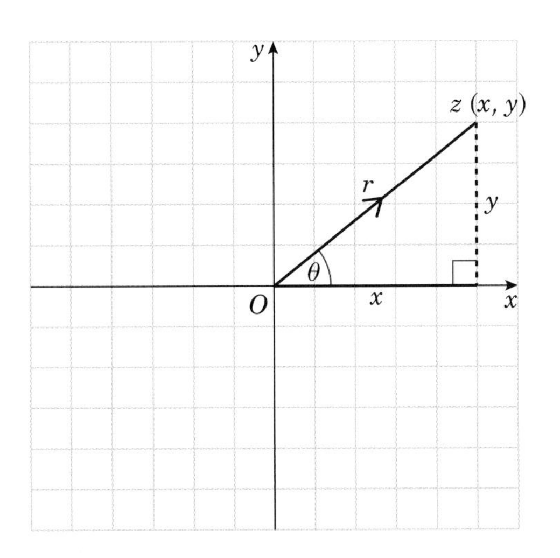

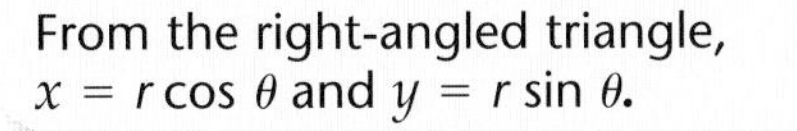
From the right-angled triangle, $x = r\cos\theta$ and $y = r\sin\theta$.

This is correct for a complex number in any of the Argand diagram quadrants.

For complex numbers z_1 and z_2, $|z_1z_2| = |z_1||z_2|$.

Here is a proof of the above result. (You do not need to remember this proof for the exam!)

Let $|z_1| = r_1$, $\arg z_1 = \theta_1$ and $|z_2| = r_2$, $\arg z_2 = \theta_2$, so

$$z_1 = r_1(\cos\theta_1 + i\sin\theta_1) \text{ and } z_2 = r_2(\cos\theta_2 + i\sin\theta_2).$$

$$z_1z_2 = r_1(\cos\theta_1 + i\sin\theta_1) \times r_2(\cos\theta_2 + i\sin\theta_2) = r_1r_2(\cos\theta_1 + i\sin\theta_1)(\cos\theta_2 + i\sin\theta_2)$$

$$= r_1r_2(\cos\theta_1\cos\theta_2 - \sin\theta_1\sin\theta_2 + i\sin\theta_1\cos\theta_2 + i\cos\theta_1\sin\theta_2)$$

$$= r_1r_2[(\cos\theta_1\cos\theta_2 - \sin\theta_1\sin\theta_2) + i(\sin\theta_1\cos\theta_2 + \cos\theta_1\sin\theta_2)]$$

But $(\cos\theta_1\cos\theta_2 - \sin\theta_1\sin\theta_2) = \cos(\theta_1 + \theta_2)$ and $(\sin\theta_1\cos\theta_2 + \cos\theta_1\sin\theta_2) = \sin(\theta_1 + \theta_2)$

So $z_1z_2 = r_1r_2[\cos(\theta_1 + \theta_2) + i\sin(\theta_1 + \theta_2)]$

You can see that this gives z_1z_2 in modulus-argument form, with $|z_1z_2| = r_1r_2$.

So $|z_1z_2| = r_1r_2 = |z_1||z_2|$

(Also, in fact, $\arg(z_1z_2) = \theta_1 + \theta_2$)

Example 27

a Express the numbers $z_1 = 1 + i\sqrt{3}$ and $z_2 = -3 - 3i$ in the form $r(\cos\theta + i\sin\theta)$.

b Write down the value of $|z_1 z_2|$.

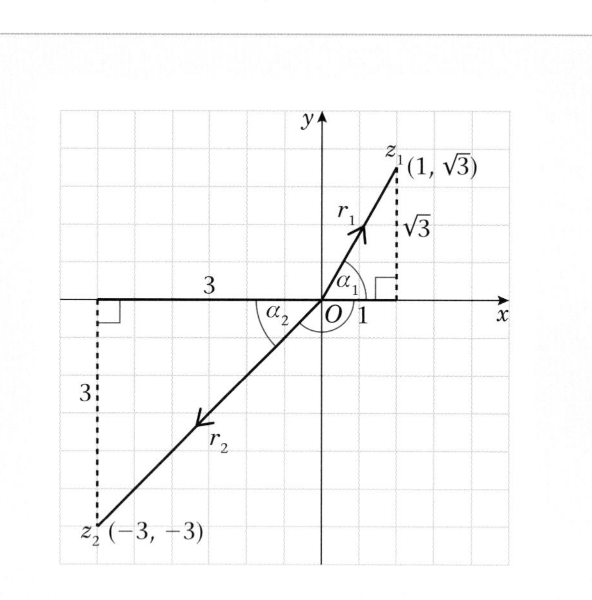

Sketch the Argand diagram, showing the position of the numbers.

Modulus: $r_1 = |z_1| = |1 + i\sqrt{3}| = \sqrt{1^2 + (\sqrt{3})^2} = \sqrt{4} = 2$

Argument: $\tan\alpha_1 = \frac{\sqrt{3}}{1} = \sqrt{3}$ $\quad \alpha_1 = \frac{\pi}{3}$

$\theta_1 = \arg z_1 = \frac{\pi}{3}$

z_1 is in the first quadrant, so this angle is the required argument (measured anticlockwise from the positive real axis).

Modulus: $r_2 = |z_2| = |-3 - 3i|$

$= \sqrt{(-3)^2 + (-3)^2}$

$= \sqrt{18} = \sqrt{9}\sqrt{2}$

$= 3\sqrt{2}$

Argument: $\tan\alpha_2 = \frac{3}{3} = 1$ $\quad \alpha_2 = \frac{\pi}{4}$

$\theta_2 = \arg z_2 = -\left(\pi - \frac{\pi}{4}\right) = -\frac{3\pi}{4}$

z_2 is in the third quadrant, so the required argument is $-(\pi - \alpha_2)$ (clockwise from the positive real axis is negative).

So $\quad z_1 = 2\left(\cos\frac{\pi}{3} + i\sin\frac{\pi}{3}\right)$

and $\quad z_2 = 3\sqrt{2}\left(\cos\left(-\frac{3\pi}{4}\right) + i\sin\left(-\frac{3\pi}{4}\right)\right)$

Using $|z_1 z_2| = r_1 r_2 = |z_1||z_2|$, $|z_1 z_2| = r_1 r_2 = 2 \times 3\sqrt{2} = 6\sqrt{2}$

Exercise 1F

1 Express these in the form $r(\cos\theta + \text{i}\sin\theta)$, giving exact values of r and θ where possible, or values to two decimal places otherwise.

a $2 + 2\text{i}$

b 3i

c $-3 + 4\text{i}$

d $1 - \sqrt{3}\text{i}$

e $-2 - 5\text{i}$

f -20

g $7 - 24\text{i}$

h $-5 + 5\text{i}$

2 Express these in the form $r(\cos\theta + \text{i}\sin\theta)$, giving exact values of r and θ where possible, or values to two decimal places otherwise.

a $\dfrac{3}{1 + \text{i}\sqrt{3}}$

b $\dfrac{1}{2 - \text{i}}$

c $\dfrac{1 + \text{i}}{1 - \text{i}}$

3 Write in the form $a + \text{i}b$, where $a \in \mathbb{R}$ and $b \in \mathbb{R}$.

a $3\sqrt{2}\left(\cos\frac{\pi}{4} + \text{i}\sin\frac{\pi}{4}\right)$

b $6\left(\cos\frac{3\pi}{4} + \text{i}\sin\frac{3\pi}{4}\right)$

c $\sqrt{3}\left(\cos\frac{\pi}{3} + \text{i}\sin\frac{\pi}{3}\right)$

d $7\left(\cos\left(-\frac{\pi}{2}\right) + \text{i}\sin\left(-\frac{\pi}{2}\right)\right)$

e $4\left(\cos\left(-\frac{5\pi}{6}\right) + \text{i}\sin\left(-\frac{5\pi}{6}\right)\right)$

4 In each case, find $|z_1|$, $|z_2|$ and z_1z_2, and verify that $|z_1z_2| = |z_1||z_2|$.

a $z_1 = 3 + 4\text{i}$ $\quad z_2 = 4 - 3\text{i}$

b $z_1 = -1 + 2\text{i}$ $\quad z_2 = 4 + 2\text{i}$

c $z_1 = 5 + 12\text{i}$ $\quad z_2 = 7 + 24\text{i}$

d $z_1 = \sqrt{3} + \text{i}\sqrt{2}$ $\quad z_2 = -\sqrt{2} + \text{i}\sqrt{3}$

1.7 You can solve problems involving complex numbers.

- **You can solve problems by equating real parts and imaginary parts from each side of an equation involving complex numbers.**
- **This technique can be used to find the square roots of a complex number.**
- **If $x_1 + \text{i}y_1 = x_2 + \text{i}y_2$, then $x_1 = x_2$ and $y_1 = y_2$.**

Example 28

Given that $3 + 5\text{i} = (a + \text{i}b)(1 + \text{i})$, where a and b are real, find the value of a and the value of b.

$(a + \text{i}b)(1 + \text{i}) = a(1 + \text{i}) + \text{i}b(1 + \text{i})$

$= a + a\text{i} + b\text{i} - b$

$= (a - b) + \text{i}(a + b)$

So $(a - b) + \text{i}(a + b) = 3 + 5\text{i}$

i $a - b = 3$ — Equate the real parts from each side of the equation.

ii $a + b = 5$ — Equate the imaginary parts from each side of the equation.

Adding **i** and **ii**: $2a = 8$

$a = 4$

Substituting into equation **i**: — Solve equations **i** and **ii** simultaneously.

$4 - b = 3$

$b = 1$

Example 29

Find the square roots of $3 + 4\text{i}$.

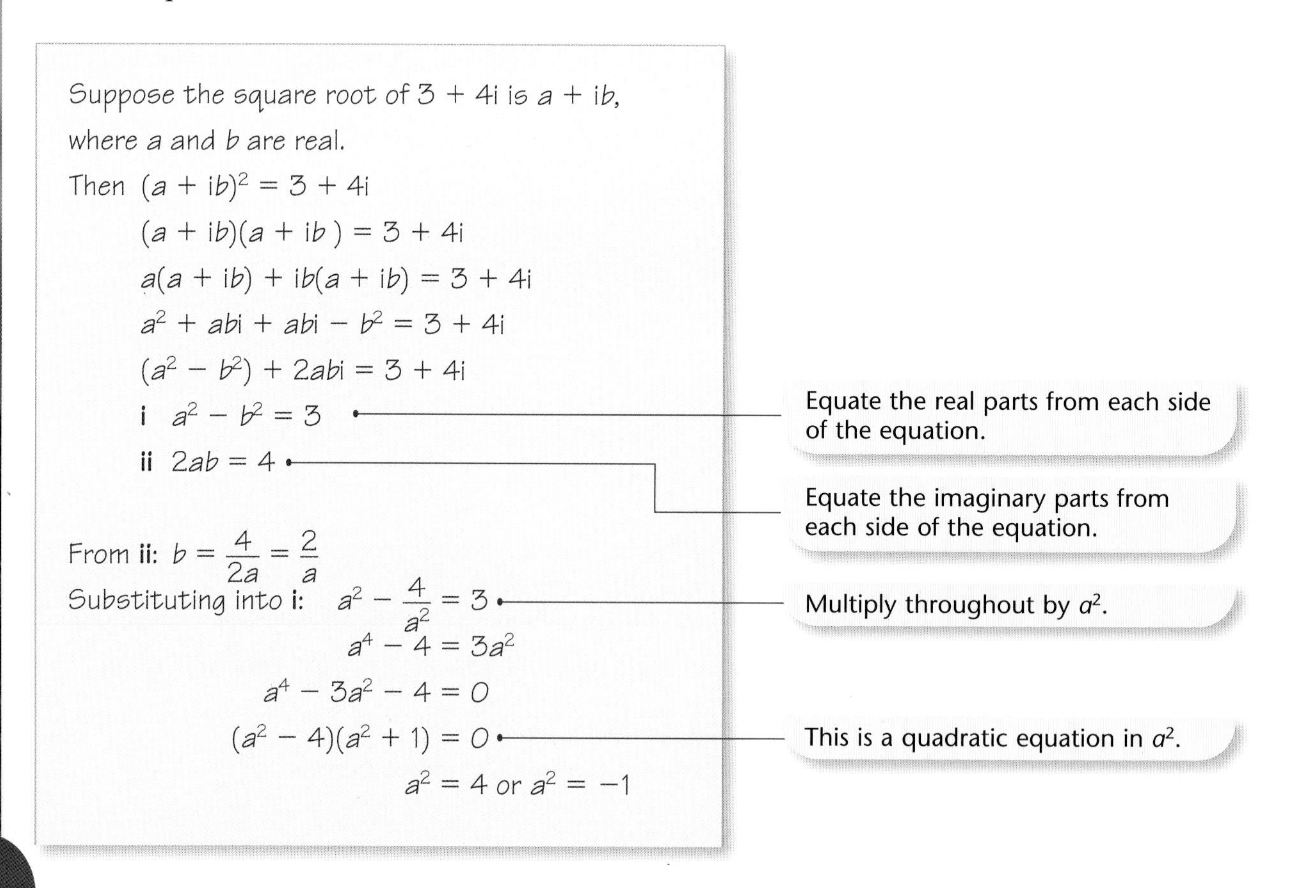

Suppose the square root of $3 + 4\text{i}$ is $a + \text{i}b$, where a and b are real.

Then $(a + \text{i}b)^2 = 3 + 4\text{i}$

$(a + \text{i}b)(a + \text{i}b) = 3 + 4\text{i}$

$a(a + \text{i}b) + \text{i}b(a + \text{i}b) = 3 + 4\text{i}$

$a^2 + ab\text{i} + ab\text{i} - b^2 = 3 + 4\text{i}$

$(a^2 - b^2) + 2ab\text{i} = 3 + 4\text{i}$

i $a^2 - b^2 = 3$ — Equate the real parts from each side of the equation.

ii $2ab = 4$ — Equate the imaginary parts from each side of the equation.

From **ii**: $b = \frac{4}{2a} = \frac{2}{a}$

Substituting into **i**: $a^2 - \frac{4}{a^2} = 3$ — Multiply throughout by a^2.

$a^4 - 4 = 3a^2$

$a^4 - 3a^2 - 4 = 0$

$(a^2 - 4)(a^2 + 1) = 0$ — This is a quadratic equation in a^2.

$a^2 = 4$ or $a^2 = -1$

Since a is real, $a^2 = -1$ has no solutions.

Solutions are $a = 2$ or $a = -2$.

Substituting back into $b = \frac{2}{a}$:

When $a = 2$, $b = 1$

When $a = -2$, $b = -1$

So the square roots are $2 + i$ and $-2 - i$

The square roots of $3 + 4i$ are $\pm(2 + i)$.

Exercise 1G

1 $a + 2b + 2ai = 4 + 6i$, where a and b are real.
Find the value of a and the value of b.

2 $(a - b) + (a + b)i = 9 + 5i$, where a and b are real.
Find the value of a and the value of b.

3 $(a + b)(2 + i) = b + 1 + (10 + 2a)i$, where a and b are real.
Find the value of a and the value of b.

4 $(a + i)^3 = 18 + 26i$, where a is real.
Find the value of a.

5 $abi = 3a - b + 12i$, where a and b are real.
Find the value of a and the value of b.

6 Find the real numbers x and y, given that

$$\frac{1}{x + iy} = 3 - 2i$$

7 Find the real numbers x and y, given that

$$(x + iy)(1 + i) = 2 + i$$

8 Solve for real x and y

$$(x + iy)(5 - 2i) = -3 + 7i$$

Hence find the modulus and argument of $x + iy$.

9 Find the square roots of $7 + 24i$.

10 Find the square roots of $11 + 60i$.

11 Find the square roots of $5 - 12i$.

12 Find the square roots of $2i$.

1.8 You can solve some types of polynomial equations with real coefficients.

- You know that, if the roots α and β of a quadratic equation are complex, α and β are always a **complex conjugate pair.**
- Given one **complex root** of a quadratic equation, you can find the equation.
- Complex roots of a polynomial equation with real coefficients occur in **conjugate pairs.**

Example 30

7 + 2i is one of the roots of a quadratic equation with real coefficients. Find the equation.

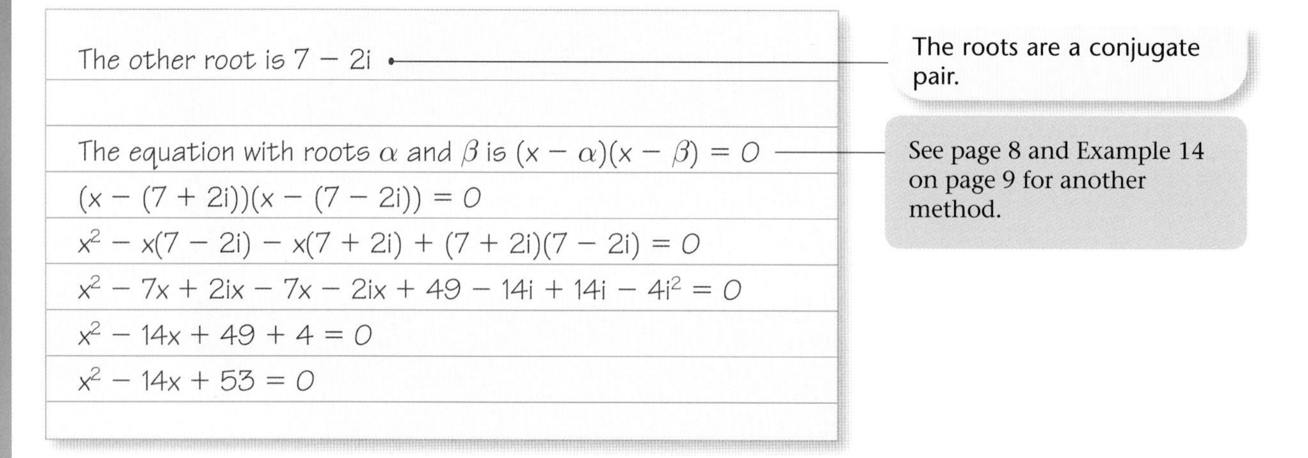

The other root is 7 − 2i

The roots are a conjugate pair.

The equation with roots α and β is $(x - \alpha)(x - \beta) = 0$

See page 8 and Example 14 on page 9 for another method.

$(x - (7 + 2i))(x - (7 - 2i)) = 0$

$x^2 - x(7 - 2i) - x(7 + 2i) + (7 + 2i)(7 - 2i) = 0$

$x^2 - 7x + 2ix - 7x - 2ix + 49 - 14i + 14i - 4i^2 = 0$

$x^2 - 14x + 49 + 4 = 0$

$x^2 - 14x + 53 = 0$

- An equation of the form $ax^3 + bx^2 + cx + d = 0$ is called a **cubic equation**, and has three roots.

Example 31

Show that $x = 2$ is a solution of the cubic equation $x^3 - 6x^2 + 21x - 26 = 0$.
Hence solve the equation completely.

For $x = 2$, $x^3 - 6x^2 + 21x - 26 = 8 - 24 + 42 - 26 = 0$

So $x = 2$ is a solution of the equation, so $x - 2$ is a factor of $x^3 - 6x^2 + 21x - 26$

$$
\begin{array}{r}
x^2 - 4x + 13 \\
x - 2 \overline{) x^3 - 6x^2 + 21x - 26} \\
\underline{x^3 - 2x^2} \qquad\qquad\quad \\
-4x^2 + 21x \qquad \\
\underline{-4x^2 + 8x} \qquad \\
13x - 26 \\
\underline{13x - 26} \\
0
\end{array}
$$

Use long division (or another method) to find the quadratic factor.

$x^3 - 6x^2 + 21x - 26 = (x - 2)(x^2 - 4x + 13) = 0$

Solving $x^2 - 4x + 13 = 0$

$x^2 - 4x = (x - 2)^2 - 4$

$x^2 - 4x + 13 = (x - 2)^2 - 4 + 13 = (x - 2)^2 + 9$

$(x - 2)^2 + 9 = 0$

$(x - 2)^2 = -9$

$x - 2 = \pm\sqrt{(-9)} = \pm 3i$

$x = 2 \pm 3i$

$x = 2 + 3i, x = 2 - 3i$

So the 3 roots of the equation are 2, 2 + 3i, and 2 − 3i.

The other two roots are found by solving the quadratic equation.

Solve by completing the square. Alternatively, you could use the quadratic formula.

The quadratic equation has complex roots, which must be a conjugate pair.

Note that, for a cubic equation,
either **i** all three roots are real,
or **ii** one root is real and the other two roots form a complex conjugate pair.

Example 32

Given that −1 is a root of the equation $x^3 - x^2 + 3x + k = 0$,

a find the value of k,

b find the other two roots of the equation.

a If −1 is a root,

$(-1)^3 - (-1)^2 + 3(-1) + k = 0$

$-1 - 1 - 3 + k = 0$

$k = 5$

b −1 is a root of the equation, so $x + 1$ is a factor of $x^3 - x^2 + 3x + 5$.

$$\begin{array}{r} x^2 - 2x + 5 \\ x + 1 \overline{) x^3 - x^2 + 3x + 5} \\ \underline{x^3 + x^2} \qquad\qquad \\ -2x^2 + 3x \qquad \\ \underline{-2x^2 - 2x} \qquad \\ 5x + 5 \\ \underline{5x + 5} \\ 0 \end{array}$$

Use long division (or another method) to find the quadratic factor.

$x^3 - x^2 + 3x + 5 = (x + 1)(x^2 - 2x + 5) = 0$

Solving $x^2 - 2x + 5 = 0$

The other two roots are found by solving the quadratic equation.

$x^2 - 2x = (x - 1)^2 - 1$

$x^2 - 2x + 5 = (x - 1)^2 - 1 + 5 = (x - 1)^2 + 4$

Solve by completing the square. Alternatively, you could use the quadratic formula.

$(x - 1)^2 + 4 = 0$

$(x - 1)^2 = -4$

$x - 1 = \pm\sqrt{(-4)} = \pm 2i$

$x = 1 \pm 2i$

The quadratic equation has complex roots, which must be a conjugate pair.

$x = 1 + 2i, x = 1 - 2i$

So the other two roots of the equation are $1 + 2i$ and $1 - 2i$.

- **An equation of the form $ax^4 + bx^3 + cx^2 + dx + e = 0$ is called a quartic equation, and has four roots.**

Example 33

Given that $3 + i$ is a root of the quartic equation $2x^4 - 3x^3 - 39x^2 + 120x - 50 = 0$, solve the equation completely.

Another root is $3 - i$.

Complex roots occur in conjugate pairs.

The equation with roots α and β is $(x - \alpha)(x - \beta) = 0$

$(x - (3 + i))(x - (3 - i)) = 0$

$x^2 - x(3 - i) - x(3 + i) + (3 + i)(3 - i) = 0$

$x^2 - 3x + ix - 3x - ix + 9 - 3i + 3i - i^2 = 0$

$x^2 - 6x + 9 + 1 = 0$

$x^2 - 6x + 10 = 0$

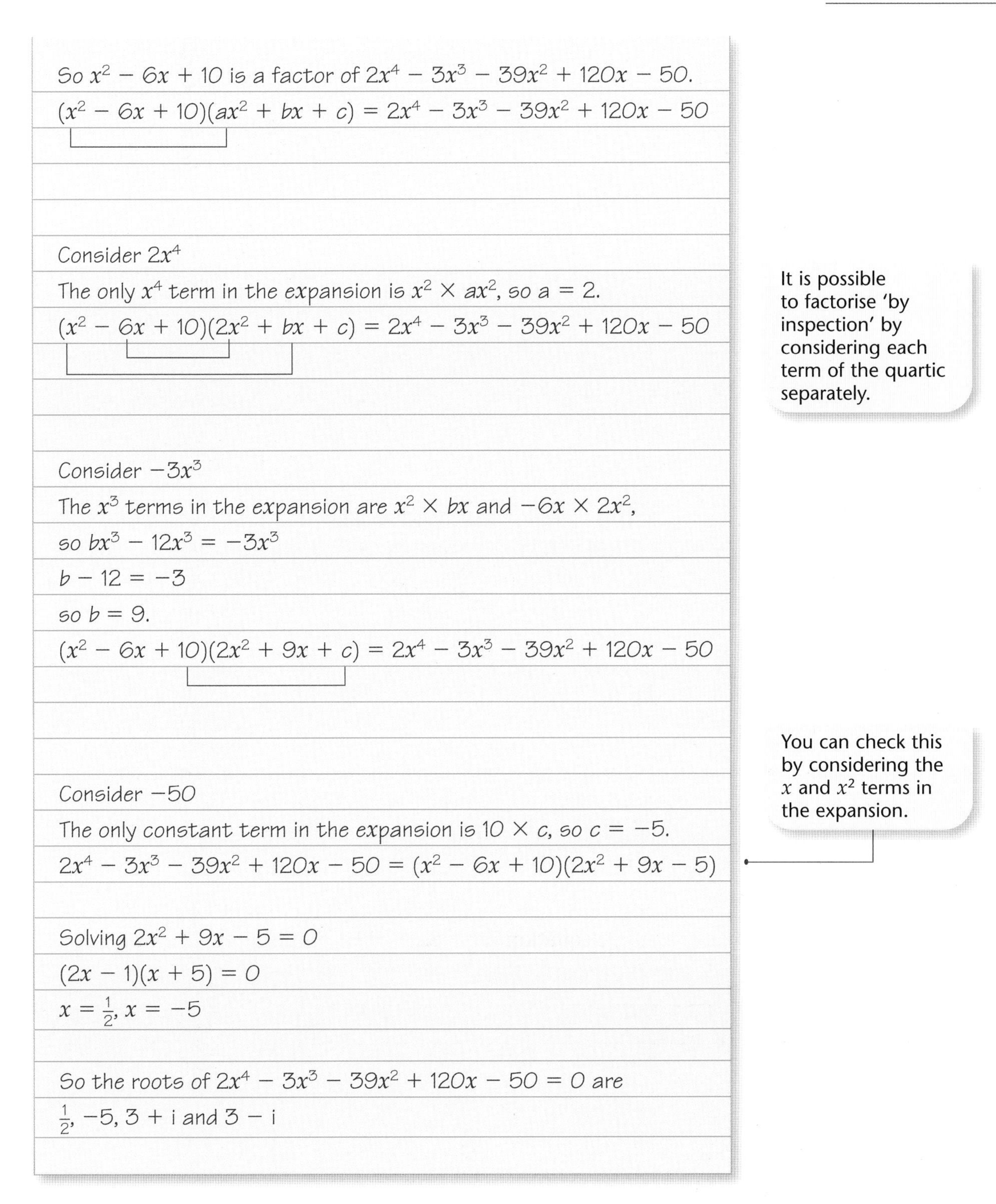

So $x^2 - 6x + 10$ is a factor of $2x^4 - 3x^3 - 39x^2 + 120x - 50$.

$(x^2 - 6x + 10)(ax^2 + bx + c) = 2x^4 - 3x^3 - 39x^2 + 120x - 50$

Consider $2x^4$

The only x^4 term in the expansion is $x^2 \times ax^2$, so $a = 2$.

$(x^2 - 6x + 10)(2x^2 + bx + c) = 2x^4 - 3x^3 - 39x^2 + 120x - 50$

It is possible to factorise 'by inspection' by considering each term of the quartic separately.

Consider $-3x^3$

The x^3 terms in the expansion are $x^2 \times bx$ and $-6x \times 2x^2$,

so $bx^3 - 12x^3 = -3x^3$

$b - 12 = -3$

so $b = 9$.

$(x^2 - 6x + 10)(2x^2 + 9x + c) = 2x^4 - 3x^3 - 39x^2 + 120x - 50$

Consider -50

The only constant term in the expansion is $10 \times c$, so $c = -5$.

$2x^4 - 3x^3 - 39x^2 + 120x - 50 = (x^2 - 6x + 10)(2x^2 + 9x - 5)$

You can check this by considering the x and x^2 terms in the expansion.

Solving $2x^2 + 9x - 5 = 0$

$(2x - 1)(x + 5) = 0$

$x = \frac{1}{2}, x = -5$

So the roots of $2x^4 - 3x^3 - 39x^2 + 120x - 50 = 0$ are

$\frac{1}{2}, -5, 3 + \mathrm{i}$ and $3 - \mathrm{i}$

Note that, for a quartic equation,

either **i** all four roots are real,

or **ii** two roots are real and the other two roots form a complex conjugate pair,

or **iii** two roots form a complex conjugate pair and the other two roots also form a complex conjugate pair.

Example 34

Show that $x^2 + 4$ is a factor of $x^4 - 2x^3 + 21x^2 - 8x + 68$.
Hence solve the equation $x^4 - 2x^3 + 21x^2 - 8x + 68 = 0$.

Using long division:

$$\begin{array}{r} x^2 - 2x + 17 \\ x^2 + 4 \overline{)\, x^4 - 2x^3 + 21x^2 - 8x + 68} \\ x^4 \quad\quad + 4x^2 \quad\quad\quad\quad \\ \hline -2x^3 + 17x^2 - 8x \quad\quad \\ -2x^3 \quad\quad\quad - 8x \quad\quad \\ \hline 17x^2 \quad\quad + 68 \\ 17x^2 \quad\quad + 68 \\ \hline 0 \end{array}$$

So $x^4 - 2x^3 + 21x^2 - 8x + 68 = (x^2 + 4)(x^2 - 2x + 17) = 0$

Either $x^2 + 4 = 0$ or $x^2 - 2x + 17 = 0$

Solving $x^2 + 4 = 0$

$x^2 = -4$

$x = \pm\sqrt{(-4)} = \pm\sqrt{(4 \times -1)} = \pm\sqrt{4}\sqrt{(-1)} = \pm 2i$

Solving $x^2 - 2x + 17 = 0$

$x^2 - 2x = (x - 1)^2 - 1$

$x^2 - 2x + 17 = (x - 1)^2 - 1 + 17 = (x - 1)^2 + 16$

$(x - 1)^2 + 16 = 0$

$(x - 1)^2 = -16$

$x - 1 = \pm\sqrt{(-16)} = \pm 4i$

$x = 1 \pm 4i$

So the roots of $x^4 - 2x^3 + 21x^2 - 8x + 68 = 0$ are

$2i, -2i, 1 + 4i$ and $1 - 4i$

It is also possible to factorise 'by inspection' by considering each term of the quartic separately, as in Example 33.

Solve by completing the square. Alternatively, you could use the quadratic formula.

Exercise 1H

1 Given that $1 + 2i$ is one of the roots of a quadratic equation with real coefficients, find the equation.

2 Given that $3 - 5i$ is one of the roots of a quadratic equation with real coefficients, find the equation.

3 Given that $a + 4i$, where a is real, is one of the roots of a quadratic equation with real coefficients, find the equation.

4 Show that $x = -1$ is a root of the equation $x^3 + 9x^2 + 33x + 25 = 0$.
Hence solve the equation completely.

5 Show that $x = 3$ is a root of the equation $2x^3 - 4x^2 - 5x - 3 = 0$.
Hence solve the equation completely.

6 Show that $x = -\frac{1}{2}$ is a root of the equation $2x^3 + 3x^2 + 3x + 1 = 0$.
Hence solve the equation completely.

7 Given that $-4 + \text{i}$ is one of the roots of the equation $x^3 + 4x^2 - 15x - 68 = 0$, solve the equation completely.

8 Given that $x^4 - 12x^3 + 31x^2 + 108x - 360 = (x^2 - 9)(x^2 + bx + c)$, find the values of b and c, and hence find all the solutions of the equation $x^4 - 12x^3 + 31x^2 + 108x - 360 = 0$.

9 Given that $2 + 3\text{i}$ is one of the roots of the equation
$x^4 + 2x^3 - x^2 + 38x + 130 = 0$, solve the equation completely.

10 Find the four roots of the equation $x^4 - 16 = 0$.
Show these roots on an Argand diagram.

11 Three of the roots of the equation $ax^5 + bx^4 + cx^3 + dx^2 + ex + f = 0$ are -2, 2i and $1 + \text{i}$.
Find the values of a, b, c, d, e and f.

Mixed exercise 1I

1 **a** Find the roots of the equation $z^2 + 2z + 17 = 0$ giving your answers in the form $a + \text{i}b$, where a and b are integers.

b Show these roots on an Argand diagram. **E**

2 $z_1 = -\text{i}$, $z_2 = 1 + \text{i}\sqrt{3}$

a Find the modulus of **i** z_1z_2 **ii** $\frac{z_1}{z_2}$.

b Find the argument of **i** z_1z_2 **ii** $\frac{z_1}{z_2}$.

Give your answers in radians as exact multiples of π. **E**

3 $z = \frac{1}{2 + \text{i}}$.

a Express in the form $a + b\text{i}$, where $a, b \in \mathbb{R}$,

i z^2 **ii** $z - \frac{1}{z}$.

b Find $|z^2|$.

c Find $\arg\left(z - \frac{1}{z}\right)$, giving your answer in degrees to one decimal place. **E**

4 The real and imaginary parts of the complex number $z = x + \text{i}y$ satisfy the equation $(2 - \text{i})\,x - (1 + 3\text{i})\,y - 7 = 0$.

a Find the value of x and the value of y.

b Find the values of **i** $|z|$ **ii** $\arg z$. **E**

5 Given that $2 + \text{i}$ is a root of the equation $z^3 - 11z + 20 = 0$, find the other roots of the equation. *E*

6 Given that $1 + 3\text{i}$ is a root of the equation $z^3 + 6z + 20 = 0$,

a find the other two roots of the equation,

b show, on a single Argand diagram, the three points representing the roots of the equation,

c prove that these three points are the vertices of a right-angled triangle. *E*

7 $z_1 = 4 + 2\text{i}$, $z_2 = -3 + \text{i}$

a Display points representing z_1 and z_2 on the same Argand diagram.

b Find the exact value of $|z_1 - z_2|$.

Given that $w = \frac{z_1}{z_2}$,

c express w in the form $a + \text{i}b$, where $a, b \in \mathbb{R}$,

d find $\arg w$, giving your answer in radians. *E*

8 Given that $3 - 2\text{i}$ is a solution of the equation

$$x^4 - 6x^3 + 19x^2 - 36x + 78 = 0,$$

a solve the equation completely,

b show on a single Argand diagram the four points that represent the roots of the equation. *E*

9 $z = \frac{a + 3\text{i}}{2 + a\text{i}}$, $\quad a \in \mathbb{R}$.

a Given that $a = 4$, find $|z|$.

b Show that there is only one value of a for which $\arg z = \frac{\pi}{4}$, and find this value. *E*

Summary of key points

1 $\sqrt{(-1)} = \text{i}$ and $\text{i}^2 = -1$.

2 An imaginary number is a number of the form $b\text{i}$, where b is a real number ($b \in \mathbb{R}$).

3 A complex number is a number of the form $a + b\text{i}$, where $a \in \mathbb{R}$ and $b \in \mathbb{R}$.

4 For the complex number $a + b\text{i}$, a is called the real part and b is called the imaginary part.

5 The complex number $z^* = a - b\text{i}$ is called the complex conjugate of the complex number $z = a + b\text{i}$.

6 If the roots α and β of a quadratic equation are complex, α and β will always be a complex conjugate pair.

7 The complex number $z = x + \text{i}y$ is represented on an Argand diagram by the point (x, y), where x and y are Cartesian coordinates.

8 The complex number $z = x + \text{i}y$ can also be represented by the vector $\overrightarrow{OP}$, where O is the origin and P is the point (x, y) on the Argand diagram.

9 Addition of complex numbers can be represented on the Argand diagram by the addition of their respective vectors on the diagram.

10 The modulus of the complex number $z = x + \text{i}y$ is given by $\sqrt{x^2 + y^2}$.

11 The modulus of the complex number $z = x + \text{i}y$ is written as r or $|z|$ or $|x + \text{i}y|$, so

$$r = \sqrt{x^2 + y^2}$$
$$|z| = \sqrt{x^2 + y^2}$$
$$|x + \text{i}y| = \sqrt{x^2 + y^2}$$

12 The modulus of any non-zero complex number is positive.

13 The argument arg z of the complex number $z = x + \text{i}y$ is the angle θ between the positive real axis and the vector representing z on the Argand diagram.

14 For the argument θ of the complex number $z = x + \text{i}y$, $\tan\theta = \frac{y}{x}$.

15 The argument θ of any complex number is such that $-\pi < \theta \leqslant \pi$ (or $-180° < \theta \leqslant 180°$). (This is sometimes referred to as the principal argument.)

16 The modulus–argument form of the complex number $z = x + \text{i}y$ is $z = r(\cos\theta + \text{i}\sin\theta)$. [$r$ is a positive real number and θ is an angle such that $-\pi < \theta \leqslant \pi$ (or $-180° < \theta \leqslant 180°$)]

17 For complex numbers z_1 and z_2, $|z_1z_2| = |z_1||z_2|$.

18 If $x_1 + \text{i}y_1 = x_2 + \text{i}y_2$, then $x_1 = x_2$ and $y_1 = y_2$.

19 An equation of the form $ax^3 + bx^2 + cx + d = 0$ is called a cubic equation, and has three roots.

20 For a cubic equation, either

a all three roots are real, or

b one root is real and the other two roots form a complex conjugate pair.

21 An equation of the form $ax^4 + bx^3 + cx^2 + dx + e = 0$ is called a quartic equation, and has four roots.

22 For a quartic equation, either

a all four roots are real, or

b two roots are real and the other two roots form a complex conjugate pair, or

c two roots form a complex conjugate pair and the other two roots also form a complex conjugate pair.

After completing this chapter you should be able to

- find approximations to the solutions of equations of the form $f(x) = 0$ using:
  - interval bisection
  - linear interpolation
  - the Newton–Raphson process.

Numerical solutions of equations

Numerical methods are used in science and engineering to help solve problems. These problems are normally modelled using computers.

The numerical methods used lead to approximate solutions to the many equations that need to be solved.

Weather forecasters use numerical methods to predict the weather, both in the immediate future (a few hours) and up to a few weeks ahead.

2.1 You can solve equations of the form $f(x) = 0$ using interval bisection.

- **If you find an interval in which $f(x)$ changes sign, then the interval must contain a root of the equation $f(x) = 0$.**
- **You then take the mid-point as the first approximation and repeat this process until you get the required accuracy.**

Example 1

Use interval bisection to find the positive root of $\sqrt{11}$ to 1 decimal place.

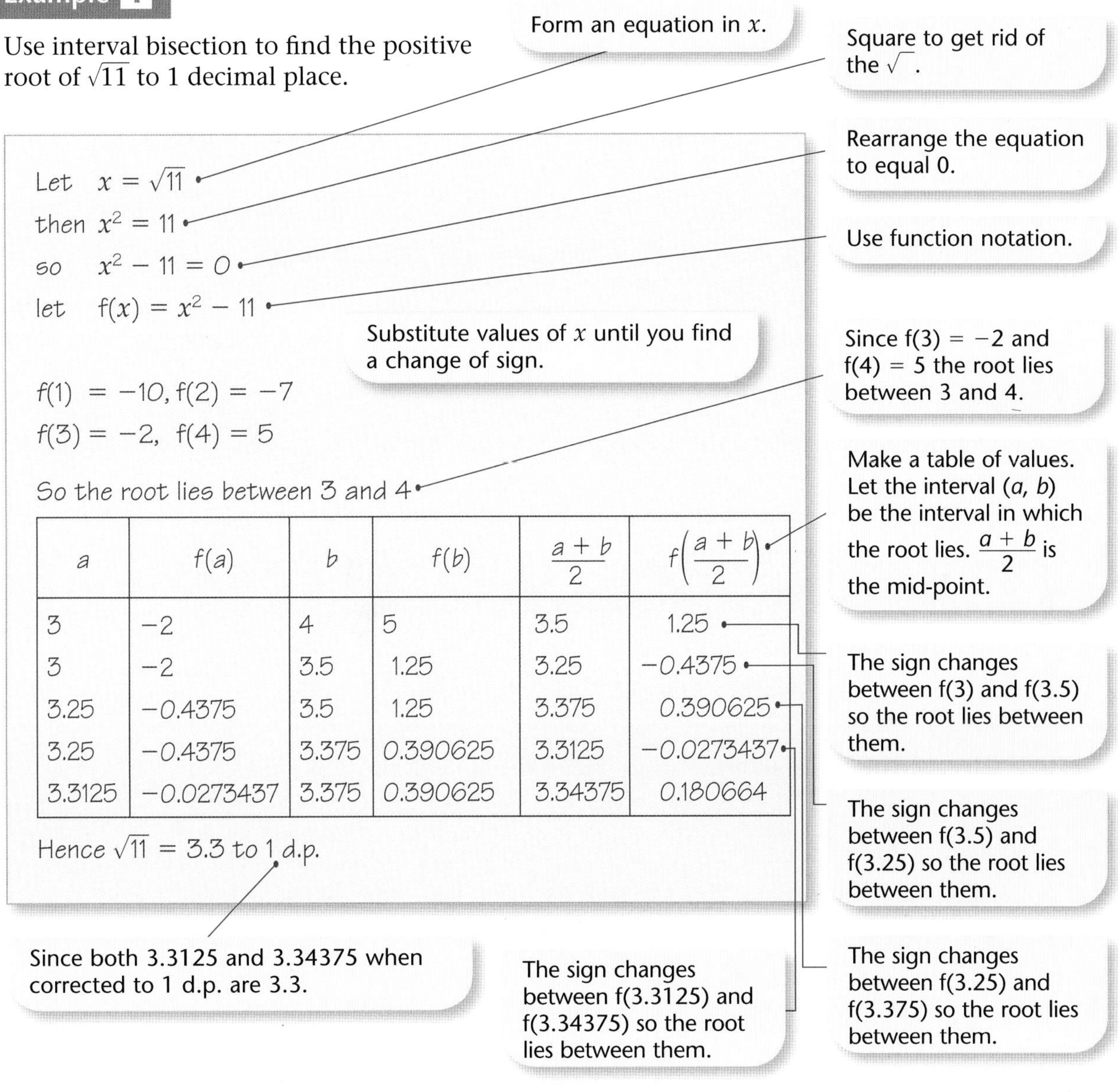

Let $x = \sqrt{11}$ — Form an equation in x.

then $x^2 = 11$ — Square to get rid of the $\sqrt{\ }$.

so $x^2 - 11 = 0$ — Rearrange the equation to equal 0.

let $f(x) = x^2 - 11$ — Use function notation.

Substitute values of x until you find a change of sign.

$f(1) = -10, f(2) = -7$

$f(3) = -2,\ f(4) = 5$

So the root lies between 3 and 4 — Since $f(3) = -2$ and $f(4) = 5$ the root lies between 3 and 4.

Make a table of values. Let the interval (a, b) be the interval in which the root lies. $\frac{a+b}{2}$ is the mid-point.

a	$f(a)$	b	$f(b)$	$\frac{a+b}{2}$	$f\left(\frac{a+b}{2}\right)$
3	−2	4	5	3.5	1.25
3	−2	3.5	1.25	3.25	−0.4375
3.25	−0.4375	3.5	1.25	3.375	0.390625
3.25	−0.4375	3.375	0.390625	3.3125	−0.0273437
3.3125	−0.0273437	3.375	0.390625	3.34375	0.180664

The sign changes between f(3) and f(3.5) so the root lies between them.

The sign changes between f(3.5) and f(3.25) so the root lies between them.

The sign changes between f(3.25) and f(3.375) so the root lies between them.

The sign changes between f(3.3125) and f(3.34375) so the root lies between them.

Hence $\sqrt{11} = 3.3$ to 1 d.p.

Since both 3.3125 and 3.34375 when corrected to 1 d.p. are 3.3.

Example 2

$$f(x) = 2^x - x - 3$$

The equation $f(x) = 0$ has a root x in the interval [2, 3].

Using the end points of this interval find by interval bisection, a first and second approximation to x.

Let $a = 2$, $b = 3$

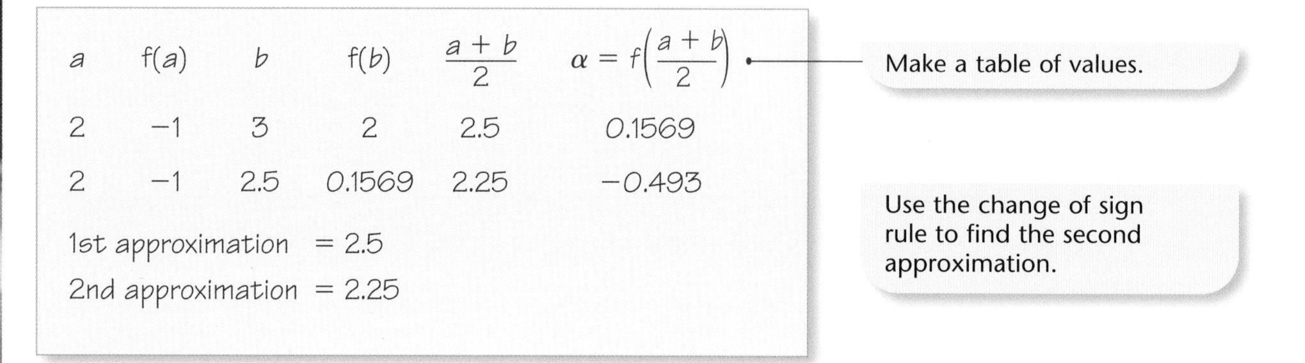

a	$f(a)$	b	$f(b)$	$\frac{a+b}{2}$	$\alpha = f\left(\frac{a+b}{2}\right)$
2	−1	3	2	2.5	0.1569
2	−1	2.5	0.1569	2.25	−0.493

1st approximation = 2.5
2nd approximation = 2.25

Make a table of values.

Use the change of sign rule to find the second approximation.

Exercise 2A

1 Use interval bisection to find the positive square root of $x^2 - 7 = 0$, correct to one decimal place.

2 **a** Show that one root of the equation $x^3 - 7x + 2 = 0$ lies in the interval [2, 3].
b Use interval bisection to find the root correct to two decimal places.

3 **a** Show that the largest positive root of the equation $0 = x^3 + 2x^2 - 8x - 3$ lies in the interval [2, 3].
b Use interval bisection to find this root correct to one decimal place.

4 **a** Show that the equation $f(x) = 1 - 2\sin x$ has one root which lies in the interval [0.5, 0.8].
b Use interval bisection four times to find this root. Give your answer correct to one decimal place.

5 **a** Show that the equation $0 = \frac{x}{2} - \frac{1}{x}$, $x > 0$, has a root in the interval [1, 2].
b Obtain the root, using interval bisection three times. Give your answer to two significant figures.

6 $f(x) = 6x - 3^x$

The equation $f(x) = 0$ has a root between $x = 2$ and $x = 3$. Starting with the interval [2, 3] use interval bisection three times to give an approximation to this root.

2.2 You can solve equations of the form $f(x) = 0$ using linear interpolation.

- **In linear interpolation, you draw a sketch of the function $f(x)$ for a given interval $[a, b]$.**
- **You then call the first approximation to the root of the function that lies in this interval x_1.**
- **You use similar triangles to find x_1.**
- **You repeat the process using an interval involving the first approximation and one of the initial limits, where there is a change of sign to find a second approximation.**
- **Repeat until you find an approximation to the required degree of accuracy.**

Example 3

a Show that the equation $x^3 + 4x - 9 = c$ has a root in the interval [1, 2].

b Use linear interpolation to find this root to one decimal place.

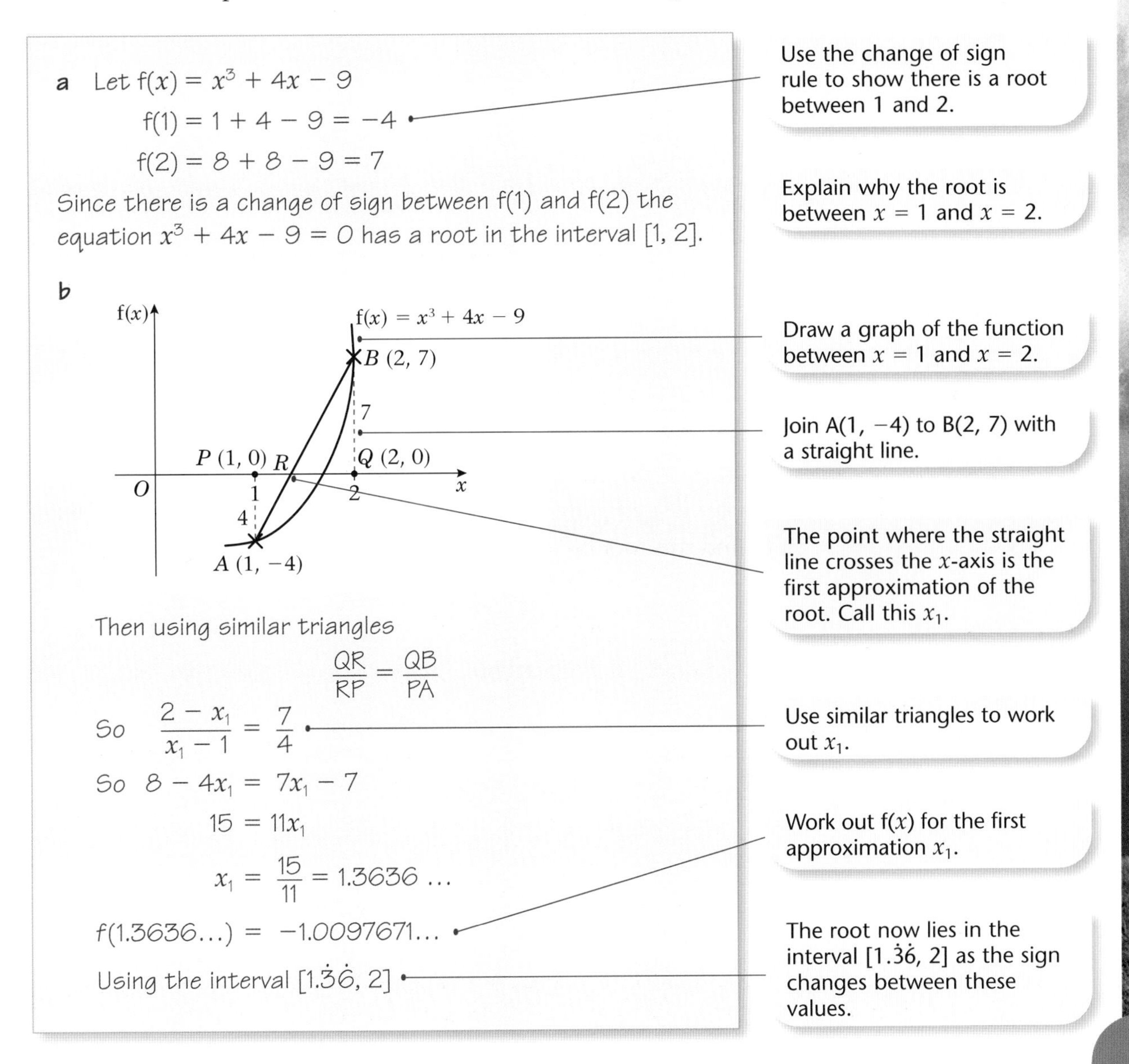

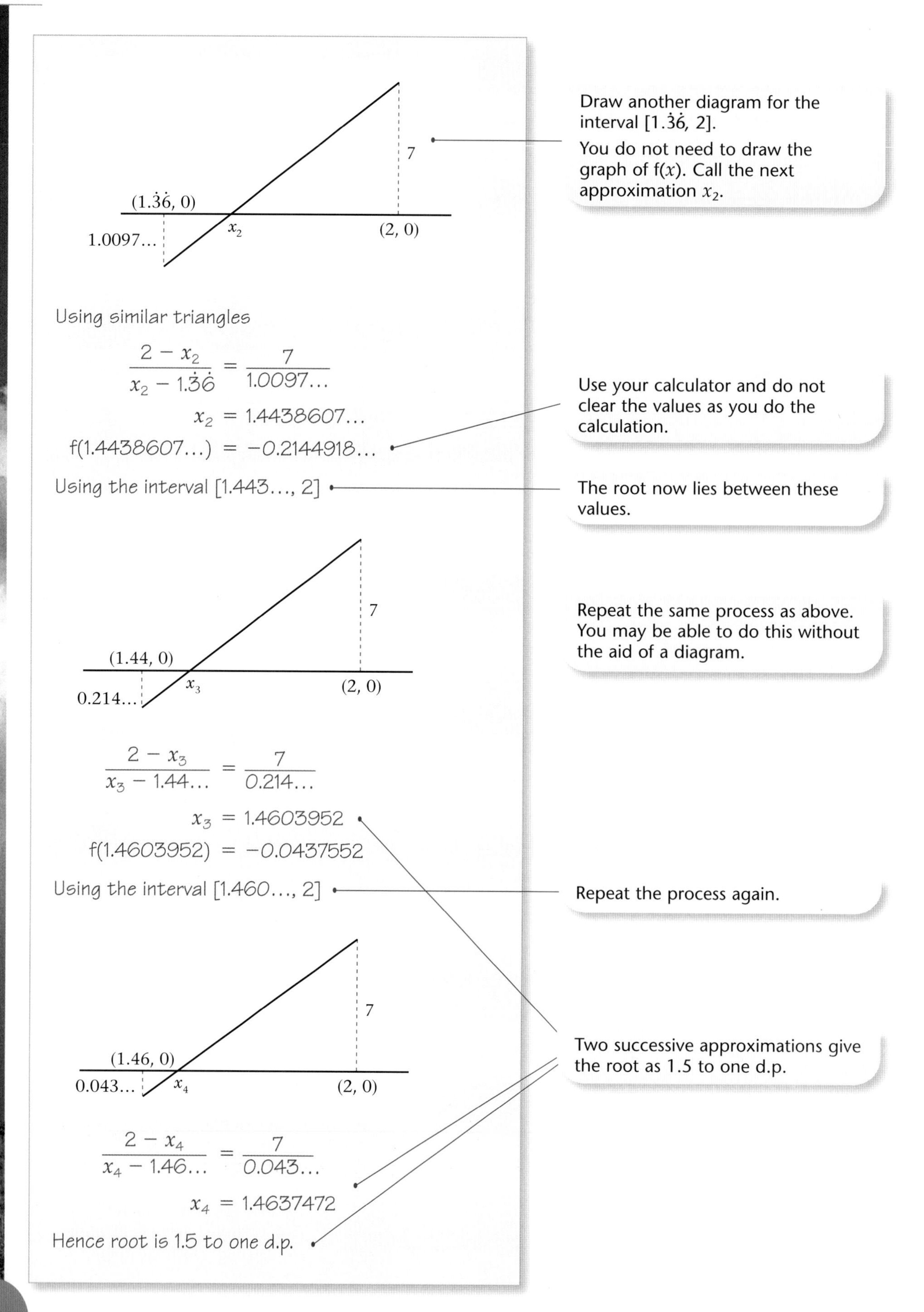

Draw another diagram for the interval $[1.\dot{3}\dot{6}, 2]$.

You do not need to draw the graph of $f(x)$. Call the next approximation x_2.

Using similar triangles

$$\frac{2 - x_2}{x_2 - 1.\dot{3}\dot{6}} = \frac{7}{1.0097\ldots}$$

$$x_2 = 1.4438607\ldots$$

$$f(1.4438607\ldots) = -0.2144918\ldots$$

Use your calculator and do not clear the values as you do the calculation.

Using the interval [1.443..., 2]

The root now lies between these values.

Repeat the same process as above. You may be able to do this without the aid of a diagram.

$$\frac{2 - x_3}{x_3 - 1.44\ldots} = \frac{7}{0.214\ldots}$$

$$x_3 = 1.4603952$$

$$f(1.4603952) = -0.0437552$$

Using the interval [1.460..., 2]

Repeat the process again.

$$\frac{2 - x_4}{x_4 - 1.46\ldots} = \frac{7}{0.043\ldots}$$

$$x_4 = 1.4637472$$

Hence root is 1.5 to one d.p.

Two successive approximations give the root as 1.5 to one d.p.

Example 4

$f(x) = 3^x - 5x$

The equation $f(x) = 0$ has a root α in this interval [2, 3].

Using the end points of this interval find, by linear interpolation, an approximation to α.

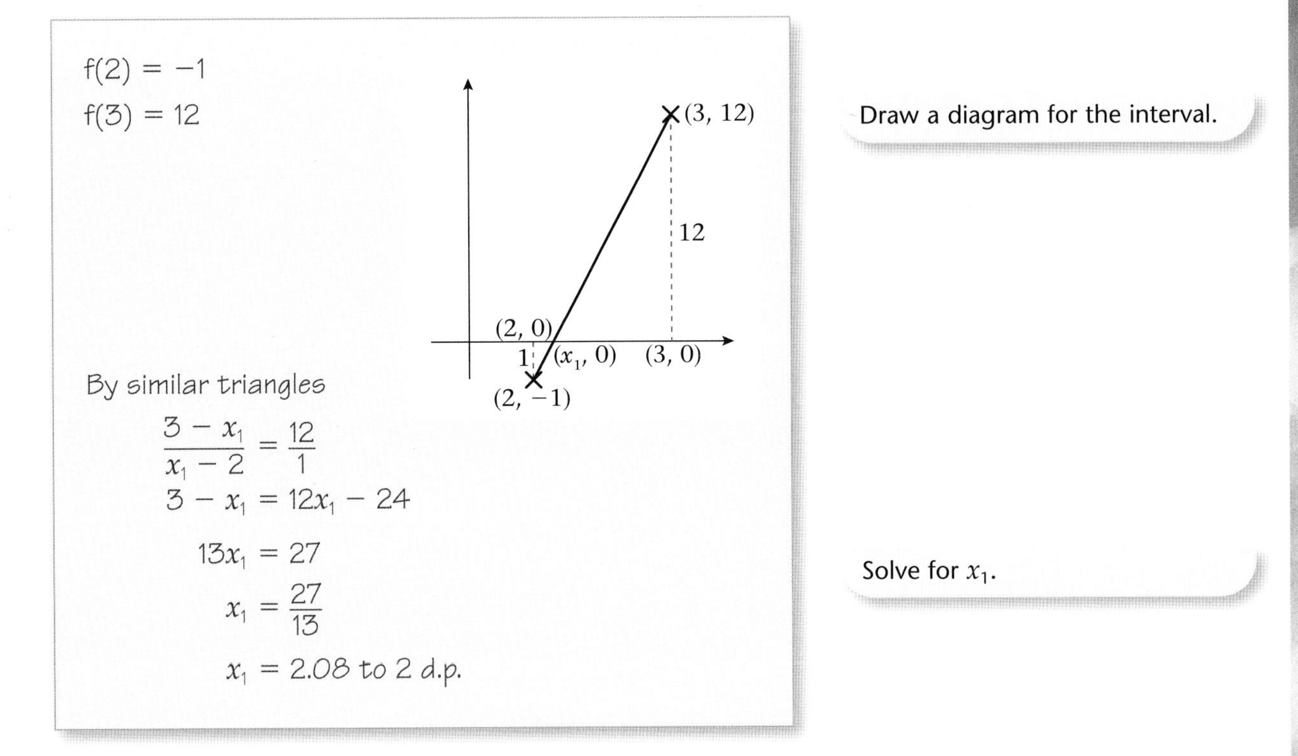

$f(2) = -1$
$f(3) = 12$

Draw a diagram for the interval.

By similar triangles

$$\frac{3 - x_1}{x_1 - 2} = \frac{12}{1}$$

$$3 - x_1 = 12x_1 - 24$$

$$13x_1 = 27$$

$$x_1 = \frac{27}{13}$$

$$x_1 = 2.08 \text{ to 2 d.p.}$$

Solve for x_1.

Exercise 2B

1 **a** Show that a root of the equation $x^3 - 3x - 5 = 0$ lies in the interval [2, 3].
b Find this root using linear interpolation correct to one decimal place.

2 **a** Show that a root of the equation $5x^3 - 8x^2 + 1 = 0$ has a root between $x = 1$ and $x = 2$.
b Find this root using linear interpolation correct to one decimal place.

3 **a** Show that a root of the equation $\frac{3}{x} + 3 = x$ lies in the interval [3, 4].
b Use linear interpolation to find this root correct to one decimal place.

4 **a** Show that a root of the equation $2x \cos x - 1 = 0$ lies in the interval [1, 1.5].
b Find this root using linear interpolation correct to two decimal places.

5 **a** Show that the largest possible root of the equation $x^3 - 2x^2 - 3 = 0$ lies in the interval [2, 3].
b Find this root correct to one decimal place using interval interpolation.

6 $f(x) = 2^x - 3x - 1$

The equation $f(x) = 0$ has a root in the interval [3, 4].

Using this interval find an approximation to x.

2.3 You can solve equations of the form $f(x) = 0$ using the Newton–Raphson process.

- **The Newton–Raphson formula is**

$$x_{n+1} = x_n - \frac{f(x_n)}{f'(x_n)}$$

- **Note that the Newton–Raphson process may not always give you a better approximation and may take you further away from a root.**

Example 5

Use the Newton–Raphson process to find the root of the equation $x^4 + x^2 = 80$ which is near to $x = 3$ correct to two decimal places

Let $f(x) = x^4 + x^2 - 80$

Then $f'(x) = 4x^3 + 2x$

Differentiate $f(x)$ to find $f'(x)$.

Let $x_0 = 3$

Then $x_1 = x_0 - \frac{f(x_0)}{f'(x_0)}$

Use the Newton–Raphson process.

$x_1 = 3 - \frac{3^4 + 3^2 - 80}{4(3)^3 + 2(3)}$

Substitute $x_0 = 3$ into the equation.

$x_1 = 3 - \frac{10}{114}$

$= 3 - 0.088 = 2.912$

Let $x_1 = 2.912$

Then $x_2 = 2.912 - \frac{2.912^4 + 2.912^2 - 80}{4(2.912)^3 + 2(2.912)}$

Repeat the process again to find x_2.

$= 2.912 - \frac{0.3858023}{104.5960581}$

$= 2.908340184$

So root $= 2.91$ to two decimal places.

As both x_1 and x_2 round to 2.91.

Example 6

$\mathrm{f}(x) = x^3 + 2x^2 - 5x - 4$

a Use differentiation to find $\mathrm{f}'(x)$.

The equation $\mathrm{f}(x) = 0$ has a root α in the interval [1, 2]

b Using 2 as a first approximation to α, use the Newton–Raphson process twice to find an approximation for α. Give your answer correct to three decimal places.

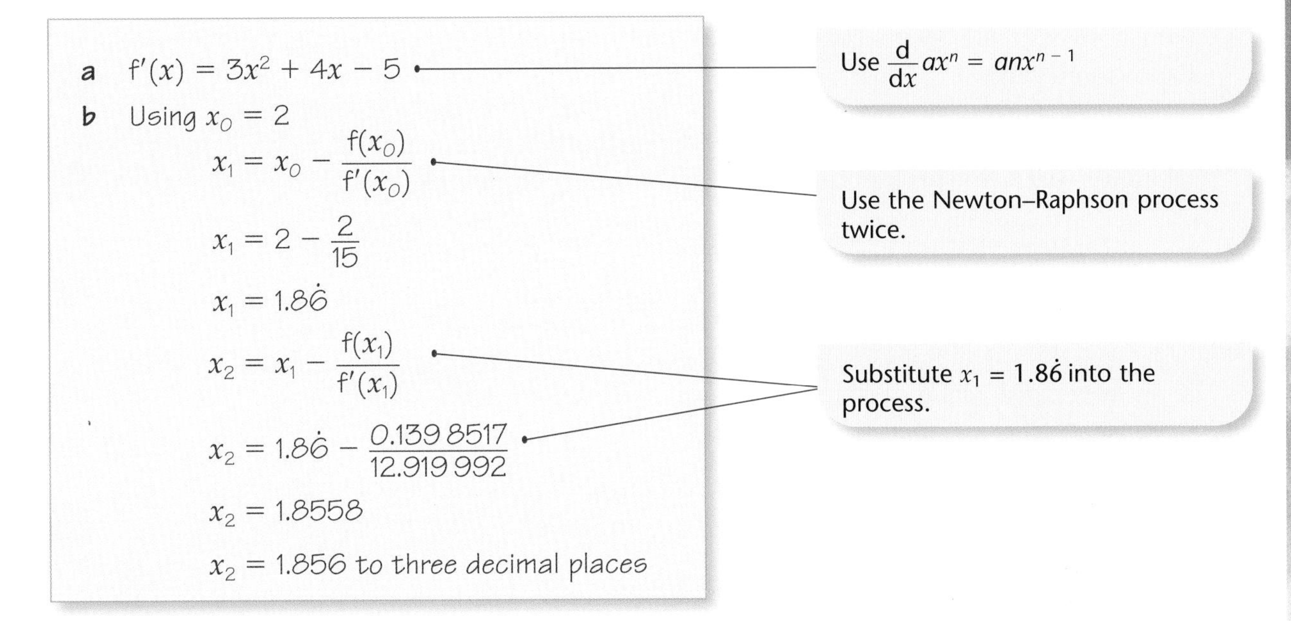

Exercise 2C

1 Show that the equation $x^3 - 2x - 1 = 0$ has a root between 1 and 2. Find the root correct to two decimal places using the Newton–Raphson process.

2 Use the Newton–Raphson process to find the positive root of the equation $x^3 + 2x^2 - 6x - 3 = 0$ correct to two decimal places.

3 Find the smallest positive root of the equation $x^4 + x^2 - 80 = 0$ correct to two decimal places. Use the Newton–Raphson process.

4 Apply the Newton–Raphson process to find the negative root of the equation $x^3 - 5x + 2 = 0$ correct to two decimal places.

5 Show that the equation $2x^3 - 4x^2 - 1 = 0$ has a root in the interval [2, 3]. Taking 3 as a first approximation to this root, use the Newton–Raphson process to find this root correct to two decimal places.

6 $\mathrm{f}(x) = x^3 - 3x^2 + 5x - 4$

Taking 1.4 as a first approximation to a root, x, of this equation, use Newton–Raphson process once to obtain a second approximation to x. Give your answer to three decimal places.

7 Use the Newton–Raphson process twice to find the root of the equation $2x^3 + 5x = 70$ which is near to $x = 3$. Give your answer to three decimal places.

Mixed exercise 2D

1 Given that $f(x) = x^3 - 2x + 2$ has a root in the interval $[-1, -2]$, use interval bisection on the interval $[-1, -2]$ to obtain the root correct to one decimal place.

2 Show that the equation $x^3 - 12x - 7.2 = 0$ has one positive and two negative roots. Obtain the positive root correct to three significant figures using the Newton–Raphson process. *E*

3 Find, correct to one decimal place, the real root of $x^3 + 2x - 1 = 0$ by using the Newton–Raphson process. *E*

4 Use the Newton–Raphson process to find the real root of the equation $x^3 + 2x^2 + 4x - 6 = 0$, taking $x = 0.9$ as the first approximation and carrying out one iteration. *E*

5 Use linear interpolation to find the positive root of the equation $x^3 - 5x + 3 = 0$ correct to one decimal place. *E*

6 $f(x) = x^3 + x^2 - 6$.

a Show that the real root of $f(x) = 0$ lies in the interval $[1, 2]$.

b Use the linear interpolation on the interval $[1, 2]$ to find the first approximation to x.

c Use the Newton–Raphson process on $f(x)$ once, starting with your answer to **b**, to find another approximation to x, giving your answer correct to two decimal places. *E*

7 The equation $\cos x = \frac{1}{4}x$ has a root in the interval $[1.0, 1.4]$. Use linear interpolation once in the interval $[1.0, 1.4]$ to find an estimate of the root, giving your answer correct to two decimal places.

8 $f(x) = x^3 - 3x - 6$

Use the Newton–Raphson process to find the positive root of this equation correct to two decimal places.

Summary of key points

1 You can solve equations of the form $f(x) = 0$ using interval bisection.

If you find an interval in which $f(x)$ changes sign, then the interval must contain a root of the equation $f(x) = 0$. You then take the mid-point as the first approximation and repeat this process until you get your required accuracy

2 You can solve equations of the form $f(x) = 0$ using linear interpolation.

3 You can solve equations of the form $f(x) = 0$ using the Newton–Raphson process.

4 The Newton–Raphson formula is

$$x_{n+1} = x_n - \frac{f(x_n)}{f'(x_n)}$$

5 The Newton–Raphson process may not always give you a better approximation and may take you further away from the root.

After completing this chapter you should be able to:

- plot and sketch a curve expressed parametrically
- work with the Cartesian equation and parametric equations of
  - a parabola
  - a rectangular hyperbola
- understand the focus–directrix property of a parabola
- find the equation of the tangent and the equation of a normal to a point
  - on a parabola
  - on a rectangular hyperbola.

Coordinate systems

In this chapter you will be introduced to a parabola and its properties. You will also work with another curve called a rectangular hyperbola.

This chapter builds upon the coordinate geometry work that you have learnt in units Core 1 and Core 2. For example, you will revise material on the distance, gradient and mid-point between two points (x_1, y_1) and (x_2, y_2).

The shape of the cables found on the Clifton Suspension Bridge in Bristol is an example of the ways in which parabolas are used in the real world.

3.1 You know what parametric equations are.

Parametric equations are where the x and y coordinates of each point on a curve are expressed in the form of an independent variable, say t, which is called a **parameter**. The **parametric** equation of a curve is written in the form:

$x = \mathrm{f}(t), y = \mathrm{g}(t).$

You can define the coordinates of any point on a curve by using parametric equations.

Example 1

Sketch the curve given by the parametric equations $x = at^2$, $y = 2at$, $t \in \mathbb{R}$ where a is a positive constant.

To give an idea of the shape of the curve we choose some values for t.
Let's say $t = -3, -2, -1, 0, 1, 2, 3$.

t	-3	-2	-1	0	1	2	3
$x = at^2$	$9a$	$4a$	a	0	a	$4a$	$9a$
$y = 2at$	$-6a$	$-4a$	$-2a$	0	$2a$	$4a$	$6a$

Draw a table showing the values of t, x and y.

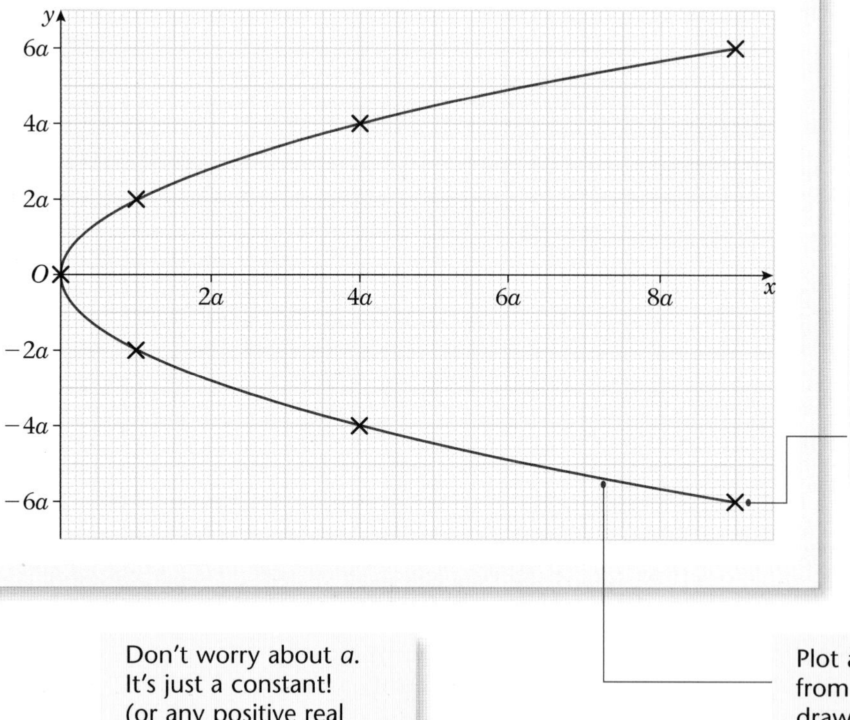

Work out the value of x and the value of y by substituting each value of t into the parametric equations $x = at^2$ and $y = 2at$.

e.g. for $t = -3$:
$x = at^2 = a(-3)^2 = 9a$
and
$y = 2at = 2a(-3)$
$= -6a$

So when $t = -3$ the curve passes through the point $(9a, -6a)$.

Don't worry about a. It's just a constant! (or any positive real number).

Plot all the (x, y) points from the table and draw a smooth curve through all the points.

Example 2

A curve has parametric equations $x = at^2$, $y = 2at$, $t \in \mathbb{R}$ where a is a positive constant. Find the Cartesian equation of the curve.

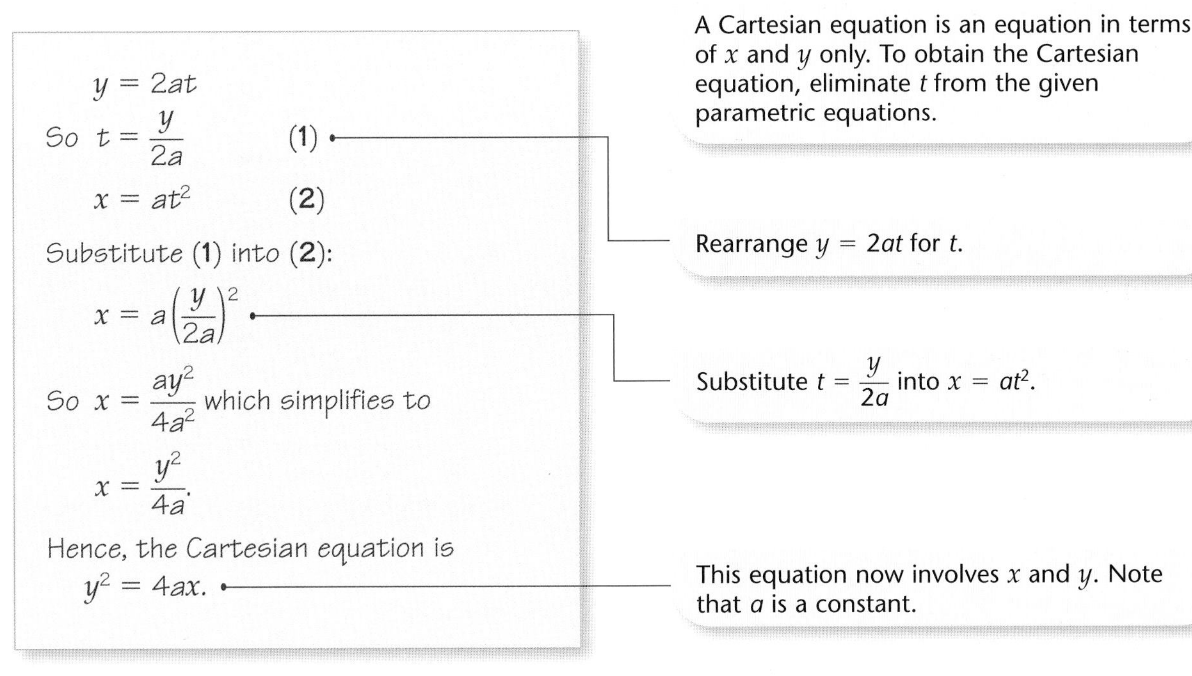

$y = 2at$

So $t = \dfrac{y}{2a}$ (1)

$x = at^2$ (2)

Substitute (1) into (2):

$x = a\left(\dfrac{y}{2a}\right)^2$

So $x = \dfrac{ay^2}{4a^2}$ which simplifies to

$x = \dfrac{y^2}{4a}$.

Hence, the Cartesian equation is

$y^2 = 4ax$.

A Cartesian equation is an equation in terms of x and y only. To obtain the Cartesian equation, eliminate t from the given parametric equations.

Rearrange $y = 2at$ for t.

Substitute $t = \dfrac{y}{2a}$ into $x = at^2$.

This equation now involves x and y. Note that a is a constant.

Example 3

A curve has parametric equations $x = ct$, $y = \frac{c}{t}$, $t \in \mathbb{R}$, $t \neq 0$ where c is a positive constant.

a Find the Cartesian equation of the curve.

b Hence sketch this curve.

a ***Method 1***

$x = ct$

So $t = \frac{x}{c}$ (1)

$y = \frac{c}{t}$ (2)

Substitute (1) into (2):

$y = \dfrac{c}{\left(\frac{x}{c}\right)}$

So $y = c \times \frac{c}{x}$

Hence, the Cartesian equation is

$y = \frac{c^2}{x}$.

To obtain the Cartesian equation, eliminate t from the given parametric equations.

Rearrange $x = ct$ for t.

Substitute $t = \frac{x}{c}$ into $y = \frac{c}{t}$.

This simplifies to $y = \frac{c^2}{x}$.

This equation now involves x and y. Note that c is a constant.

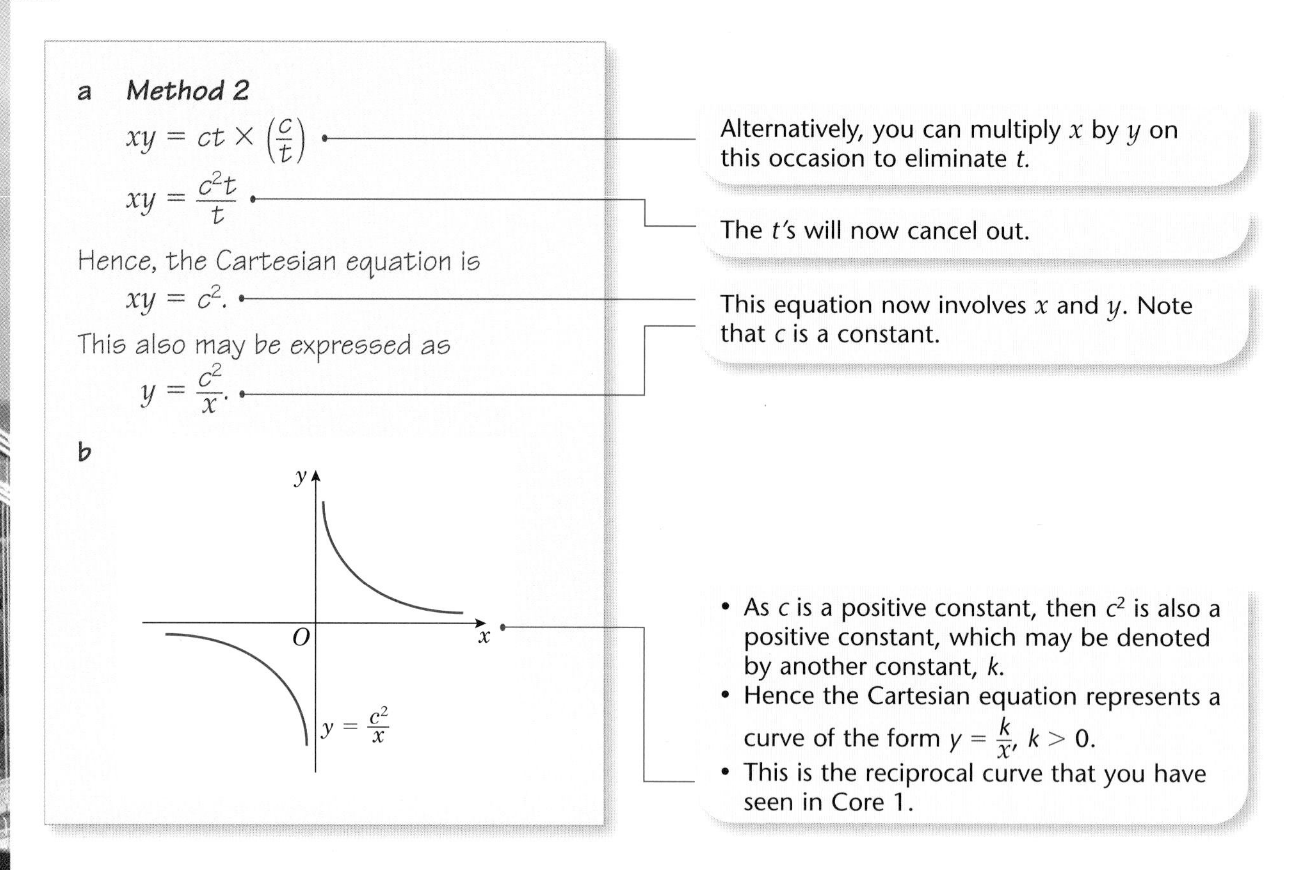

Exercise 3A

1 A curve is given by the parametric equations $x = 2t^2$, $y = 4t$. $t \in \mathbb{R}$. Copy and complete the following table and draw a graph of the curve for $-4 \leqslant t \leqslant 4$.

t	−4	−3	−2	−1	−0.5	0	0.5	1	2	3	4
$x = 2t^2$	32					0	0.5				32
$y = 4t$	−16						2				16

2 A curve is given by the parametric equations $x = 3t^2$, $y = 6t$. $t \in \mathbb{R}$. Copy and complete the following table and draw a graph of the curve for $-3 \leqslant t \leqslant 3$.

t	−3	−2	−1	−0.5	0	0.5	1	2	3
$x = 3t^2$					0				
$y = 6t$					0				

3 A curve is given by the parametric equations $x = 4t$, $y = \frac{4}{t}$. $t \in \mathbb{R}$, $t \neq 0$. Copy and complete the following table and draw a graph of the curve for $-4 \leqslant t \leqslant 4$.

t	−4	−3	−2	−1	−0.5	0.5	1	2	3	4
$x = 4t$	−16				−2					
$y = \frac{4}{t}$	−1				−8					

4 Find the Cartesian equation of the curves given by these parametric equations.

a $x = 5t^2, y = 10t$ **b** $x = \frac{1}{2}t^2, y = t$ **c** $x = 50t^2, y = 100t$

d $x = \frac{1}{5}t^2, y = \frac{2}{5}t$ **e** $x = \frac{5}{2}t^2, y = 5t$ **f** $x = \sqrt{3}t^2, y = 2\sqrt{3}t$

g $x = 4t, y = 2t^2$ **h** $x = 6t, y = 3t^2$

5 Find the Cartesian equation of the curves given by these parametric equations.

a $x = t, y = \frac{1}{t}, t \neq 0$ **b** $x = 7t, y = \frac{7}{t}, t \neq 0$

c $x = 3\sqrt{5}t, y = \frac{3\sqrt{5}}{t}, t \neq 0$ **d** $x = \frac{t}{5}, y = \frac{1}{5t}, t \neq 0$

6 A curve has parametric equations $x = 3t, y = \frac{3}{t}, t \in \mathbb{R}, t \neq 0$.

a Find the Cartesian equation of the curve.

b Hence sketch this curve.

7 A curve has parametric equations $x = \sqrt{2}t, y = \frac{\sqrt{2}}{t}, t \in \mathbb{R}, t \neq 0$.

a Find the Cartesian equation of the curve.

b Hence sketch this curve.

3.2 You know the general equation of a parabola.

- **The curve opposite is an example of a parabola which has parametric equations:**

  $x = at^2, y = 2at, t \in \mathbb{R}$,

  where a is a positive constant.

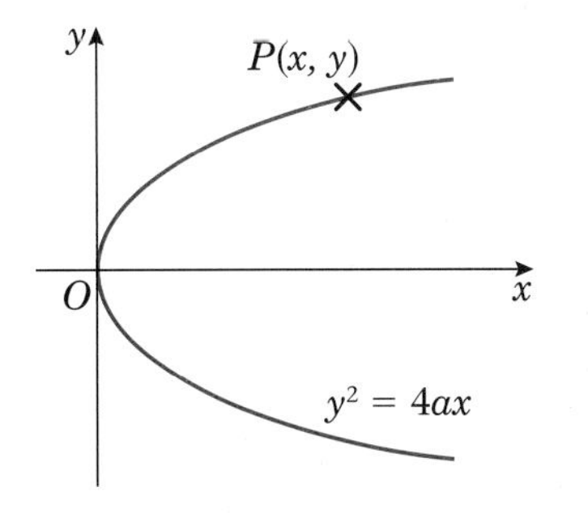

- **The Cartesian equation of this curve is $y^2 = 4ax$ where a is a positive constant.**
- **This curve is symmetrical about the x-axis.**
- **A general point P on this curve has coordinates $P(x, y)$ or $P(at^2, 2at)$.**

A *locus of points* is a set of points which obey a certain rule.

- **A parabola is the *locus of points* where every point $P(x, y)$ on the parabola is the same distance from a fixed point S, called the focus, and a fixed straight line called the directrix.**

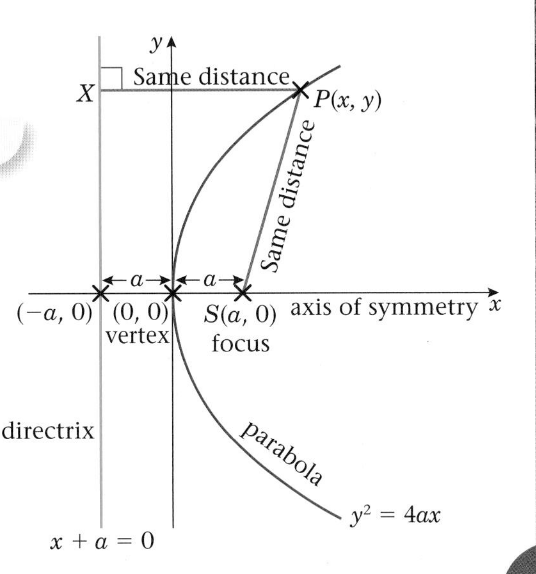

- **The parabola is the set of points where $SP = PX$.**

  The focus, S, has coordinates $(a, 0)$

  The directrix has equation $x + a = 0$.

  The vertex is at the point $(0, 0)$.

Example 4

Find an equation of the parabola with

a focus (7, 0) and directrix $x + 7 = 0$

b focus $\left(\frac{\sqrt{3}}{4}, 0\right)$ and directrix $x = -\frac{\sqrt{3}}{4}$.

a focus (7, 0) and directrix
$x + 7 = 0$

The focus and directrix are in the form $(a, 0)$ and $x + a = 0$. So $a = 7$.

So parabola has equation
$y^2 = 28x$.

Write equation in the form $y^2 = 4ax$ with $a = 7$.

b focus $\left(\frac{\sqrt{3}}{4}, 0\right)$ and directrix
$x = -\frac{\sqrt{3}}{4}$.

focus $\left(\frac{\sqrt{3}}{4}, 0\right)$ and directrix
$x + \frac{\sqrt{3}}{4} = 0$.

Write focus and directrix in the form $(a, 0)$ and $x + a = 0$.

So $a = \frac{\sqrt{3}}{4}$.

So parabola has equation
$y^2 = \sqrt{3}x$.

With $a = \frac{\sqrt{3}}{4}$, $y^2 = 4\left(\frac{\sqrt{3}}{4}\right)x$.

Example 5

Find the coordinates of the focus and an equation for the directrix of a parabola with equation

a $y^2 = 24x$,

b $y^2 = \sqrt{32}x$.

a $y^2 = 24x$

This is in the form $y^2 = 4ax$
So $4a = 24$, gives $a = \frac{24}{4} = 6$.

So the focus has coordinates (6, 0).

Focus has coordinates $(a, 0)$.

and the directrix has equation $x + 6 = 0$.

Directrix has equation $x + a = 0$.

b $y^2 = \sqrt{32}x$.

In surds, $\sqrt{32} = \sqrt{16} \times \sqrt{2} = 4\sqrt{2}$.
So $4a = 4\sqrt{2}$, gives $a = \frac{4\sqrt{2}}{4} = \sqrt{2}$.

So the focus has coordinates $(\sqrt{2}, 0)$.

Focus has coordinates $(a, 0)$.

and the directrix has equation $x + \sqrt{2} = 0$.

Directrix has equation $x + a = 0$.

Reminder

To find the distance d between two points (x_1, y_1) and (x_2, y_2) you can use the formula,

$d = \sqrt{(x_2 - x_1)^2 + (y_2 - y_1)^2}$.

This formula can also be written in the form

$d^2 = (x_2 - x_1)^2 + (y_2 - y_1)^2$.

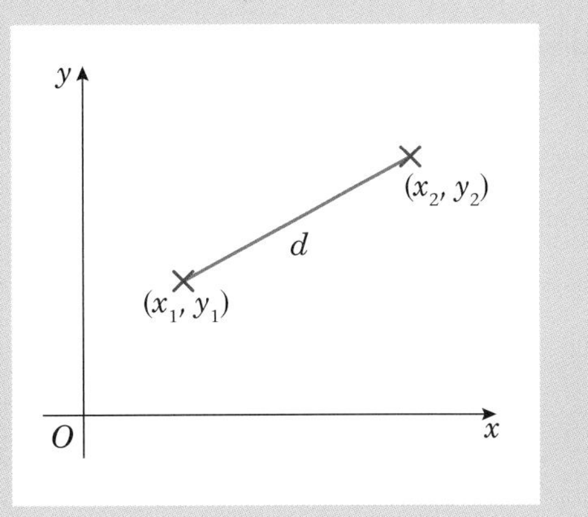

Example 6

A point $P(x, y)$ obeys a rule such that the distance of P to the point $(6, 0)$ is the same as the distance of P to the straight line $x + 6 = 0$. Prove that the locus of P has an equation of the form $y^2 = 4ax$, stating the value of the constant a.

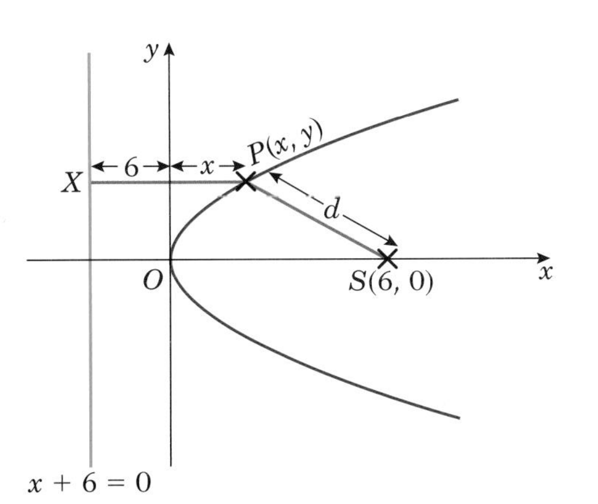

The (shortest) distance of P to the line $x + 6 = 0$ is the distance XP.

The distance SP is the same as the distance XP.

The line XP is horizontal and has distance $XP = x + 6$.

The locus of P is the curve shown.

From sketch the locus satisfies

$SP = XP$. — This means the distance SP is the same as the distance XP.

Therefore, $SP^2 = XP^2$ — Square both sides.

So, $(x - 6)^2 + (y - 0)^2 = (x - -6)^2$. — Use $d^2 = (x_2 - x_1)^2 + (y_2 - y_1)^2$ on $SP^2 = XP^2$ where $S(6, 0)$, $P(x, y)$ and $X(-6, y)$.

$x^2 - 12x + 36 + y^2 = x^2 + 12x + 36$ — Remember, $(x - 6)^2 = (x - 6)(x - 6)$

$-12x + y^2 = 12x$

which simplifies to $y^2 = 24x$.

So, the locus of P has an equation of the form $y^2 = 4ax$ where $a = 6$. — This is in the form $y^2 = 4ax$. So $4a = 24$, gives $a = \frac{24}{4} = 6$.

Exercise 3B

1 Find an equation of the parabola with

a focus (5, 0) and directrix $x + 5 = 0$,

b focus (8, 0) and directrix $x + 8 = 0$,

c focus (1, 0) and directrix $x = -1$,

d focus $\left(\frac{3}{2}, 0\right)$ and directrix $x = -\frac{3}{2}$,

e focus $\left(\frac{\sqrt{3}}{2}, 0\right)$ and directrix $x + \frac{\sqrt{3}}{2} = 0$.

2 Find the coordinates of the focus, and an equation for the directrix of a parabola with these equations.

a $y^2 = 12x$

b $y^2 = 20x$

c $y^2 = 10x$

d $y^2 = 4\sqrt{3}x$

e $y^2 = \sqrt{2}x$

f $y^2 = 5\sqrt{2}x$

3 A point $P(x, y)$ obeys a rule such that the distance of P to the point (3, 0) is the same as the distance of P to the straight line $x + 3 = 0$. Prove that the locus of P has an equation of the form $y^2 = 4ax$, stating the value of the constant a.

4 A point $P(x, y)$ obeys a rule such that the distance of P to the point $(2\sqrt{5}, 0)$ is the same as the distance of P to the straight line $x = -2\sqrt{5}$. Prove that the locus of P has an equation of the form $y^2 = 4ax$, stating the value of the constant a.

5 A point $P(x, y)$ obeys a rule such that the distance of P to the point (0, 2) is the same as the distance of P to the straight line $y = -2$.

a Prove that the locus of P has an equation of the form $y = kx^2$, stating the value of the constant k.

Given that the locus of P is a parabola,

b state the coordinates of the focus of P, and an equation of the directrix to P,

c sketch the locus of P with its focus and its directrix.

Example 7

The point $P(8, -8)$ lies on the parabola C with equation $y^2 = 8x$. The point S is the focus of the parabola. The line l passes through S and P.

a Find the coordinates of S.

b Find an equation for l, giving your answer in the form $ax + by + c = 0$, where a, b and c are integers.

The line l meets the parabola C again at the point Q. The point M is the mid-point of PQ.

c Find the coordinates of Q.

d Find the coordinates of M.

e Draw a sketch showing parabola C, the line l and the points P, Q, S and M.

a $y^2 = 8x$

The focus, S has coordinates $(2, 0)$

This is in the form $y^2 = 4ax$.
So $4a = 8$, gives $a = \frac{8}{4} = 2$.

Focus has coordinates $(a, 0)$.

b $m = \frac{-8-0}{8-2} = \frac{-8}{6}$

So $m = -\frac{4}{3}$.

l: $y - 0 = -\frac{4}{3}(x - 2)$

l: $3y = -4(x - 2)$

l: $3y = -4x + 8$

l: $4x + 3y - 8 = 0$

The line l has equation
$4x + 3y - 8 = 0$.

Use $m = \frac{y_2 - y_1}{x_2 - x_1}$, where $(x_1, y_1) = (2, 0)$ and $(x_2, y_2) = (8, -8)$.

Use $y - y_1 = m(x - x_1)$. Here $m = -\frac{4}{3}$ and $(x_1, y_1) = (2, 0)$.

Multiply both sides by 3.

Multiply out brackets.

Simplify into the form $ax + by + c = 0$.

c l: $4x + 3y - 8 = 0$ **(1)**

C: $y^2 = 8x$ **(2)**

$8x + 6y - 16 = 0$ **(3)**

$y^2 + 6y - 16 = 0$

$(y + 8)(y - 2) = 0$

$y = -8, 2.$

$4 = 8x$

$x = \frac{4}{8} = \frac{1}{2}$

The point Q has coordinates $(\frac{1}{2}, 2)$.

As the line l meets the curve C, we solve these equations simultaneously.

Multiply **(1)** by 2.

Substitute **(2)** into **(3)**.

Factorise.

$y = -8$ is at P and $y = 2$ is at Q.

Use $y^2 = 8x$ and $y = 2$ to find the x-coordinate of Q.

d The mid-point is $\left(\frac{8 + \frac{1}{2}}{2}, \frac{-8 + 2}{2}\right)$

The point M has coordinates

$\left(\frac{17}{4}, -3\right)$.

Simplify.

Use $\left(\frac{x_1 + x_2}{2}, \frac{y_1 + y_2}{2}\right)$, where
$P = (x_1, y_1) = (8, -8)$ and
$Q = (x_2, y_2) = (\frac{1}{2}, 2)$.

e

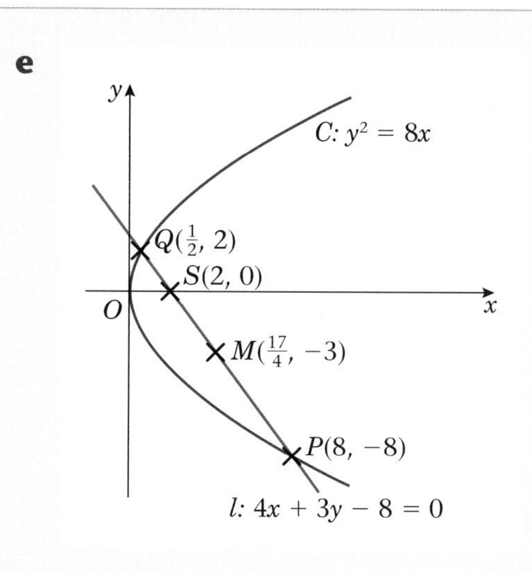

The parabola C has equation
$$y^2 = 8x$$

The line l has equation
$$4x + 3y - 8 = 0.$$

The line l cuts the parabola at the points $P(8, -8)$ and $Q(\frac{1}{2}, 2)$.

The points $S(2, 0)$ and $M(\frac{17}{4}, -3)$ also lie on the line l.

Exercise 3C

1 The line $y = 2x - 3$ meets the parabola $y^2 = 3x$ at the points P and Q. Find the coordinates of P and Q.

2 The line $y = x + 6$ meets the parabola $y^2 = 32x$ at the points A and B. Find the exact length AB giving your answer as a surd in its simplest form.

3 The line $y = x - 20$ meets the parabola $y^2 = 10x$ at the points A and B. Find the coordinates of A and B. The mid-point of AB is the point M. Find the coordinates of M.

4 The parabola C has parametric equations $x = 6t^2$, $y = 12t$. The focus to C is at the point S.

a Find a Cartesian equation of C.

b State the coordinates of S and the equation of the directrix to C.

c Sketch the graph of C.

The points P and Q on the parabola are both at a distance 9 units away from the directrix of the parabola.

d State the distance PS.

e Find the exact length PQ, giving your answer as a surd in its simplest form.

f Find the area of the triangle PQS, giving your answer in the form $k\sqrt{2}$, where k is an integer.

5 The parabola C has equation $y^2 = 4ax$, where a is a constant. The point $(\frac{5}{4}t^2, \frac{5}{2}t)$ is a general point on C.

a Find a Cartesian equation of C.

The point P lies on C with y-coordinate 5.

b Find the x-coordinate of P.

The point Q lies on the directrix of C where $y = 3$. The line l passes through the points P and Q.

c Find the coordinates of Q.

d Find an equation for l, giving your answer in the form $ax + by + c = 0$, where a, b and c are integers.

6 A parabola C has equation $y^2 = 4x$. The point S is the focus to C.

a Find the coordinates of S.

The point P with y-coordinate 4 lies on C.

b Find the x-coordinate of P.

The line l passes through S and P.

c Find an equation for l, giving your answer in the form $ax + by + c = 0$, where a, b and c are integers.

The line l meets C again at the point Q.

d Find the coordinates of Q.

e Find the distance of the directrix of C to the point Q.

7 The diagram shows the point P which lies on the parabola C with equation $y^2 = 12x$.

The point S is the focus of C. The points Q and R lie on the directrix to C. The line segment QP is parallel to the line segment RS as shown in the diagram. The distance of PS is 12 units.

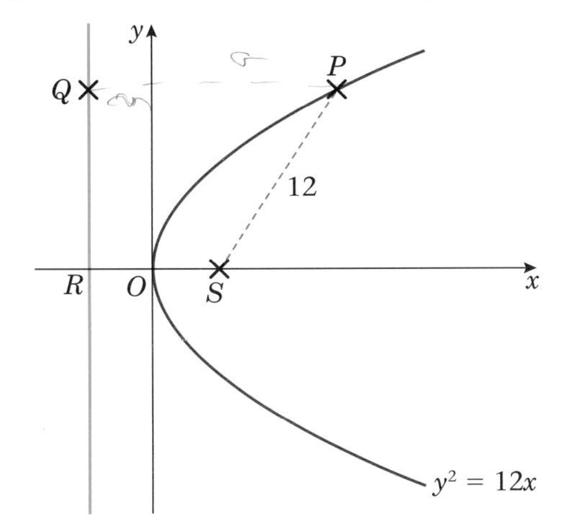

a Find the coordinates of R and S.

b Hence find the exact coordinates of P and Q.

c Find the area of the quadrilateral $PQRS$, giving your answer in the form $k\sqrt{3}$, where k is an integer.

8 The points $P(16, 8)$ and $Q(4, b)$, where $b < 0$ lie on the parabola C with equation $y^2 = 4ax$.

a Find the values of a and b.

P and Q also lie on the line l. The mid-point of PQ is the point R.

b Find an equation of l, giving your answer in the form $y = mx + c$, where m and c are constants to be determined.

c Find the coordinates of R.

The line n is perpendicular to l and passes through R.

d Find an equation of n, giving your answer in the form $y = mx + c$, where m and c are constants to be determined.

The line n meets the parabola C at two points.

e Show that the x-coordinates of these two points can be written in the form $x = \lambda \pm \mu\sqrt{13}$, where λ and μ are integers to be determined.

3.3 You know an equation for a rectangular hyperbola and can find tangents and normals.

- The curve opposite is an example of a **rectangular hyperbola** which has parametric equations:

$$x = ct,\ y = \frac{c}{t},\ t \in \mathbb{R},\ t \neq 0$$

where c is a positive constant.

- The Cartesian equation of this curve is $xy = c^2$, where c is a positive constant.

- The curve has asymptotes with equations $x = 0$ (the y-axis) and $y = 0$ (the x-axis).

- A general point P on this curve has coordinates $P(x, y)$ or $P\left(ct, \frac{c}{t}\right)$.

Example 8

The point P, where $x = 2$, lies on the rectangular hyperbola H with equation $xy = 8$.

Find

a the equation of the tangent **T**,

b the equation of the normal **N**,

to H at the point P, giving your answers in the form $ax + by + c = 0$, where a, b and c are integers.

a H: $xy = 8$

$y = \frac{8}{x} \Rightarrow y = 8x^{-1}$ — Rearrange the equation for H in the form $y = x^n$.

$\frac{dy}{dx} = -8x^{-2} = -\frac{8}{x^2}$ — Differentiate to determine the gradient of H and therefore the gradient of the tangent to H.

When $x = 2$, $m_T = \frac{dy}{dx} = -\frac{8}{2^2} = -2$ — Substitute $x = 2$ to calculate the gradient of the tangent to H.

When $x = 2$, $y = \frac{8}{2} = 4$ — Find the y-coordinate when $x = 2$. Hence P has coordinates $(2, 4)$.

T: $y - 4 = -2(x - 2)$ — Use $y - y_1 = m_T(x - x_1)$ to find the equation of the tangent, T. Here $m_T = -2$ and $(x_1, y_1) = (2, 4)$.

T: $2x + y - 8 = 0$ — Then rearrange into the required form.

Therefore, the equation of the tangent to H at P is

$2x + y - 8 = 0$.

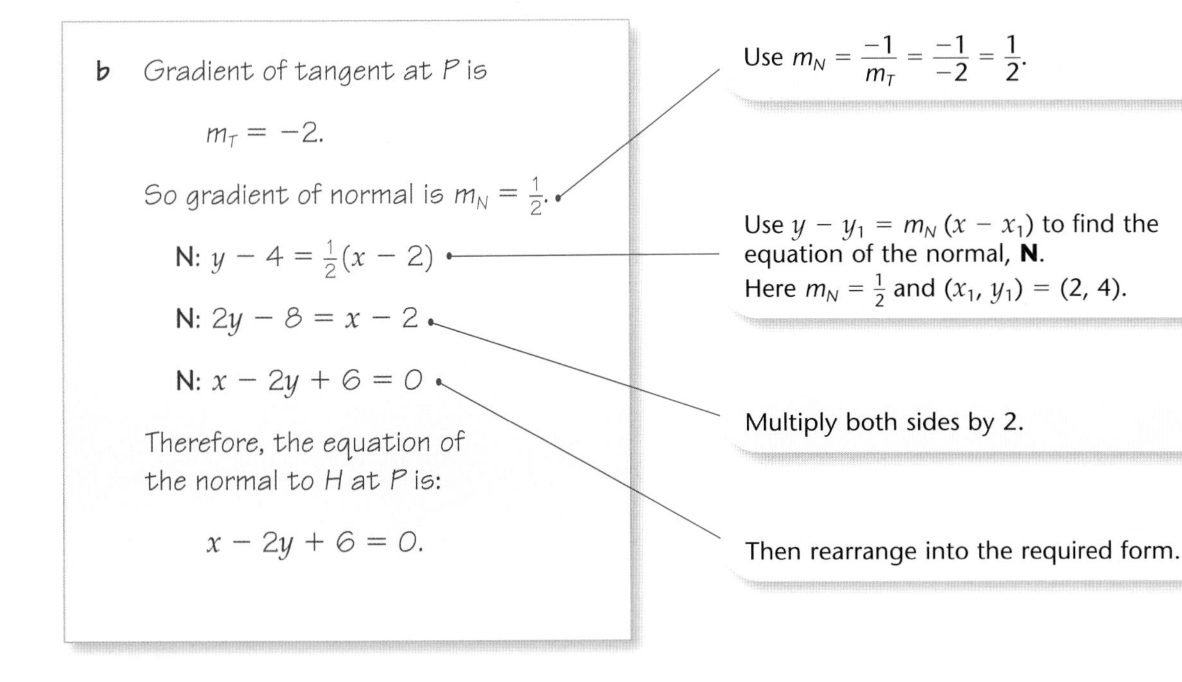

b Gradient of tangent at P is

$m_T = -2$.

So gradient of normal is $m_N = \frac{1}{2}$.

N: $y - 4 = \frac{1}{2}(x - 2)$

N: $2y - 8 = x - 2$

N: $x - 2y + 6 = 0$

Therefore, the equation of the normal to H at P is:

$x - 2y + 6 = 0$.

Use $m_N = \frac{-1}{m_T} = \frac{-1}{-2} = \frac{1}{2}$.

Use $y - y_1 = m_N(x - x_1)$ to find the equation of the normal, **N**. Here $m_N = \frac{1}{2}$ and $(x_1, y_1) = (2, 4)$.

Multiply both sides by 2.

Then rearrange into the required form.

Example 9

The distinct points A and B, where $x = 3$ lie on the parabola C with equation $y^2 = 27x$. The line l_1 is the tangent to C at A and the line l_2 is the tangent to C at B. Given that at A, $y > 0$,

a find the coordinates of A and B.

b Draw a sketch showing the parabola C. Indicate on your sketch the points A and B and the lines l_1 and l_2.

c Find:

i an equation for l_1,

ii an equation for l_2,

giving your answers in the form $ax + by + c = 0$, where a, b and c are integers.

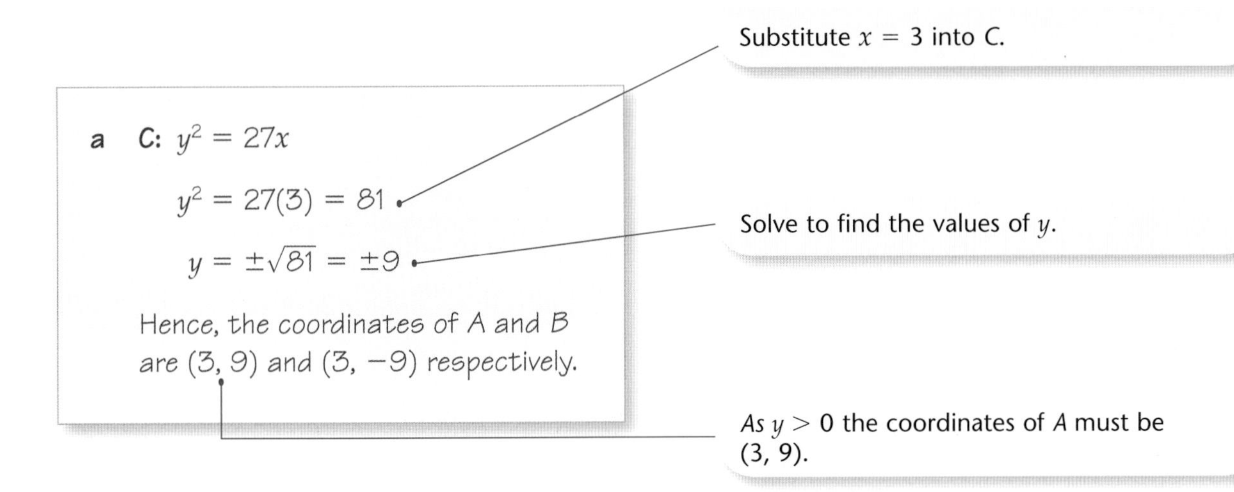

Substitute $x = 3$ into C.

a C: $y^2 = 27x$

$y^2 = 27(3) = 81$

$y = \pm\sqrt{81} = \pm 9$

Hence, the coordinates of A and B are $(3, 9)$ and $(3, -9)$ respectively.

Solve to find the values of y.

As $y > 0$ the coordinates of A must be $(3, 9)$.

b The tangent line l_1 has a positive gradient.

Hence, at A, the gradient of the curve $\frac{dy}{dx}$ is positive.

The equation of the curve for $y > 0$ is $y = +\sqrt{27}x^{\frac{1}{2}}$ or $y = +3\sqrt{3}x^{\frac{1}{2}}$.

The equation of the curve for $y < 0$ is $y = -\sqrt{27}x^{\frac{1}{2}}$ or $y = -3\sqrt{3}x^{\frac{1}{2}}$.

The tangent line l_2 has a negative gradient.

Hence, at B, the gradient of the curve $\frac{dy}{dx}$ is negative.

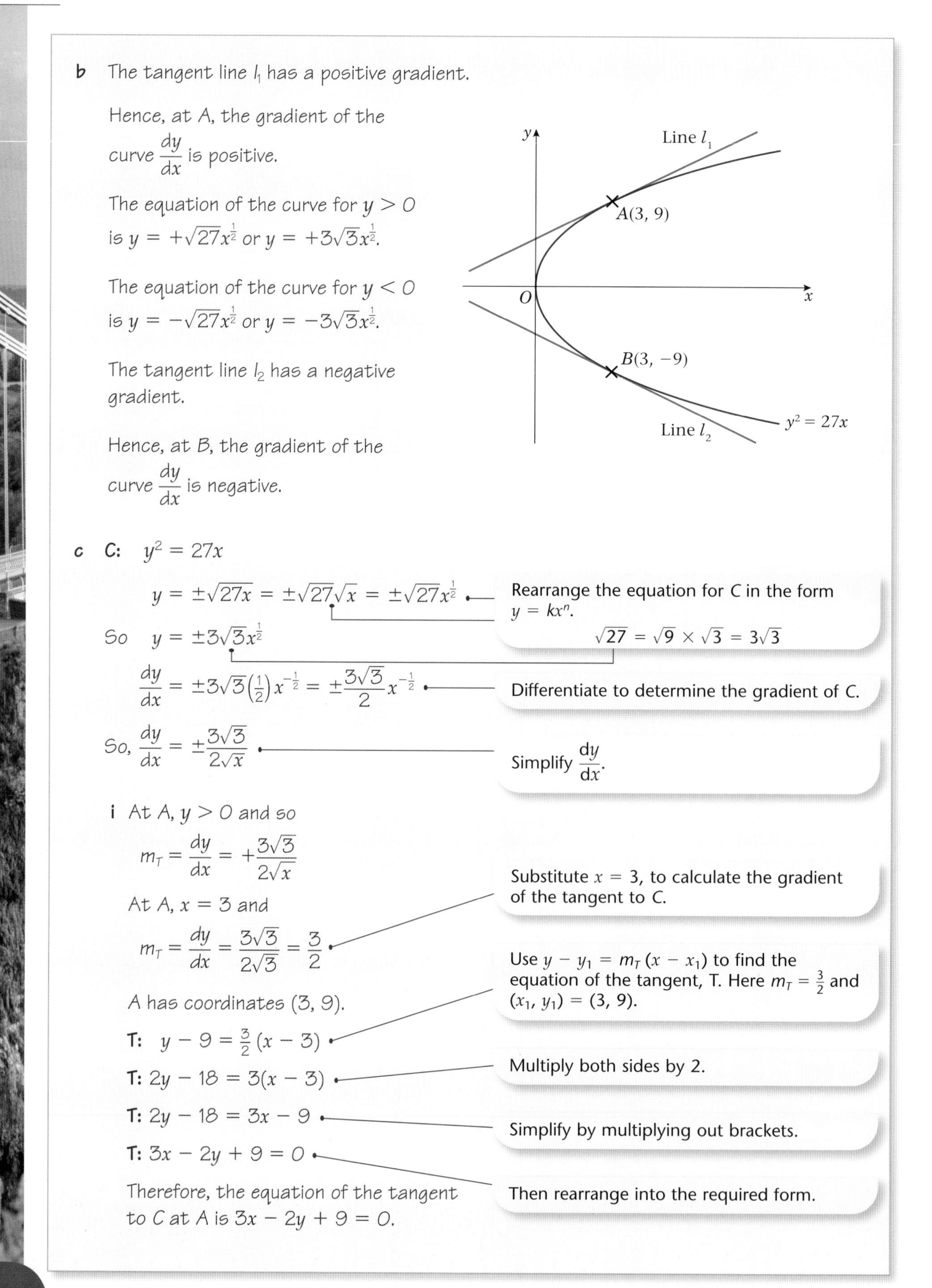

c C: $y^2 = 27x$

$$y = \pm\sqrt{27x} = \pm\sqrt{27}\sqrt{x} = \pm\sqrt{27}x^{\frac{1}{2}}$$

Rearrange the equation for C in the form $y = kx^n$.

$\sqrt{27} = \sqrt{9} \times \sqrt{3} = 3\sqrt{3}$

So $y = \pm 3\sqrt{3}x^{\frac{1}{2}}$

$$\frac{dy}{dx} = \pm 3\sqrt{3}\left(\tfrac{1}{2}\right)x^{-\frac{1}{2}} = \pm\frac{3\sqrt{3}}{2}x^{-\frac{1}{2}}$$

Differentiate to determine the gradient of C.

So, $\frac{dy}{dx} = \pm\frac{3\sqrt{3}}{2\sqrt{x}}$

Simplify $\frac{dy}{dx}$.

i At A, $y > 0$ and so

$$m_T = \frac{dy}{dx} = +\frac{3\sqrt{3}}{2\sqrt{x}}$$

At A, $x = 3$ and

$$m_T = \frac{dy}{dx} = \frac{3\sqrt{3}}{2\sqrt{3}} = \frac{3}{2}$$

Substitute $x = 3$, to calculate the gradient of the tangent to C.

A has coordinates (3, 9).

T: $y - 9 = \frac{3}{2}(x - 3)$

Use $y - y_1 = m_T(x - x_1)$ to find the equation of the tangent, T. Here $m_T = \frac{3}{2}$ and $(x_1, y_1) = (3, 9)$.

T: $2y - 18 = 3(x - 3)$

Multiply both sides by 2.

T: $2y - 18 = 3x - 9$

Simplify by multiplying out brackets.

T: $3x - 2y + 9 = 0$

Then rearrange into the required form.

Therefore, the equation of the tangent to C at A is $3x - 2y + 9 = 0$.

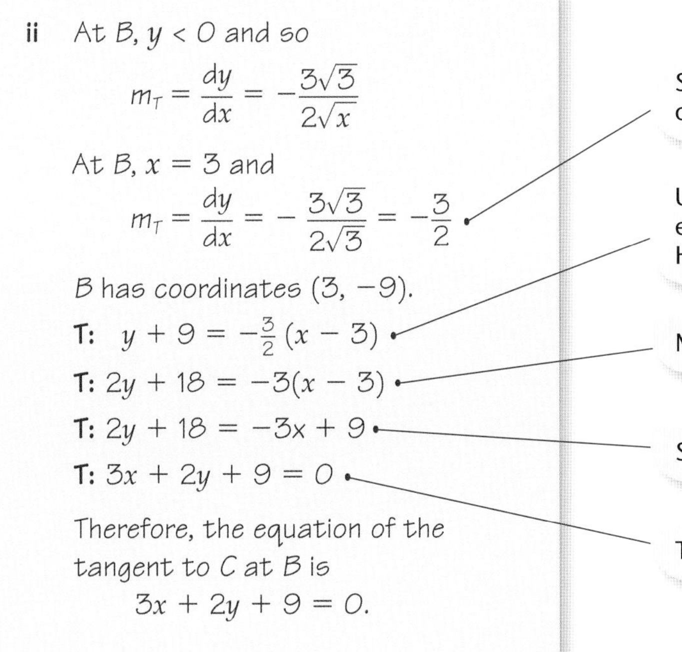

ii At B, $y < 0$ and so

$$m_T = \frac{dy}{dx} = -\frac{3\sqrt{3}}{2\sqrt{x}}$$

At B, $x = 3$ and

$$m_T = \frac{dy}{dx} = -\frac{3\sqrt{3}}{2\sqrt{3}} = -\frac{3}{2}$$

B has coordinates $(3, -9)$.

T: $y + 9 = -\frac{3}{2}(x - 3)$

T: $2y + 18 = -3(x - 3)$

T: $2y + 18 = -3x + 9$

T: $3x + 2y + 9 = 0$

Therefore, the equation of the tangent to C at B is

$$3x + 2y + 9 = 0.$$

Substitute $x = 3$, to calculate the gradient of the tangent to C.

Use $y - y_1 = m_T(x - x_1)$ to find the equation of the tangent, T.
Here $m_T = -\frac{3}{2}$ and $(x_1, y_1) = (3, -9)$.

Multiply both sides by 2.

Simplify by multiplying out brackets.

Then rearrange into the required form.

Example 10

The point P with coordinates (3, 6) lies on the parabola C with equation $y^2 = 12x$.

Find the equation of the tangent to C at P, giving your answer in the form $y = mx + c$, where m and c are constants.

C: $y^2 = 12x$

$$y = \sqrt{12x} = \sqrt{12}\sqrt{x} = \sqrt{12}x^{\frac{1}{2}}$$

So $y = 2\sqrt{3}x^{\frac{1}{2}}$

$$\frac{dy}{dx} = 2\sqrt{3}\left(\tfrac{1}{2}\right)x^{-\frac{1}{2}} = \sqrt{3}x^{-\frac{1}{2}}$$

At P, $y > 0$ and so $\dfrac{dy}{dx} = \dfrac{\sqrt{3}}{\sqrt{x}}$

When $x = 3$, $m_T = \dfrac{dy}{dx} = \dfrac{\sqrt{3}}{\sqrt{3}} = 1$

P has coordinates (3, 6).

T: $y - 6 = 1(x - 3)$

T: $y = x + 3$

Therefore, the equation of the tangent to C at P is $y = x + 3$.

At P, $y > 0$ and so you only need to take the positive square root for y. Rearrange the equation for C in the form $y = kx^n$.

$$\sqrt{12} = \sqrt{4} \times \sqrt{3} = 2\sqrt{3}$$

Differentiate to determine the gradient of C.

Simplify $\dfrac{dy}{dx}$.

Substitute $x = 3$, to calculate the gradient of the tangent to C.

Use $y - y_1 = m_T(x - x_1)$ to find the equation of the tangent, T. Here $m_T = 1$ and $(x_1, y_1) = (3, 6)$.

Then rearrange into the required form.

In Examples 8, 9 and 10 it is possible for you to find the gradient of a parabola or rectangular hyperbola by parametric differentiation or implicit differentiation. You will learn parametric differentiation and implicit differentiation in unit Core 4.

Exercise 3D

1 Find the equation of the tangent to the curve

a $y^2 = 4x$ at the point (16, 8)

b $y^2 = 8x$ at the point $(4, 4\sqrt{2})$

c $xy = 25$ at the point (5, 5)

d $xy = 4$ at the point where $x = \frac{1}{2}$

e $y^2 = 7x$ at the point $(7, -7)$

f $xy = 16$ at the point where $x = 2\sqrt{2}$.

Give your answers in the form $ax + by + c = 0$.

2 Find the equation of the normal to the curve

a $y^2 = 20x$ at the point where $y = 10$,

b $xy = 9$ at the point $\left(-\frac{3}{2}, -6\right)$.

Give your answers in the form $ax + by + c = 0$, where a, b and c are integers.

3 The point $P(4, 8)$ lies on the parabola with equation $y^2 = 4ax$. Find

a the value of a,

b an equation of the normal to C at P.

The normal to C at P cuts the parabola again at the point Q. Find

c the coordinates of Q,

d the length PQ, giving your answer as a simplified surd.

4 The point $A(-2, -16)$ lies on the rectangular hyperbola H with equation $xy = 32$.

a Find an equation of the normal to H at A.

The normal to H at A meets H again at the point B.

b Find the coordinates of B.

5 The points $P(4, 12)$ and $Q(-8, -6)$ lie on the rectangular hyperbola H with equation $xy = 48$.

a Show that an equation of the line PQ is $3x - 2y + 12 = 0$.

The point A lies on H. The normal to H at A is parallel to the chord PQ.

b Find the exact coordinates of the two possible positions of A.

6 The curve H is defined by the equations $x = \sqrt{3}t$, $y = \frac{\sqrt{3}}{t}$, $t \in \mathbb{R}$, $t \neq 0$.
The point P lies on H with x-coordinate $2\sqrt{3}$. Find:

a a Cartesian equation for the curve H,

b an equation of the normal to H at P.

The normal to H at P meets H again at the point Q.

c Find the exact coordinates of Q.

7 The point $P(4t^2, 8t)$ lies on the parabola C with equation $y^2 = 16x$. The point P also lies on the rectangular hyperbola H with equation $xy = 4$.

a Find the value of t, and hence find the coordinates of P.

The normal to H at P meets the x-axis at the point N.

b Find the coordinates of N.

The tangent to C at P meets the x-axis at the point T.

c Find the coordinates of T.

d Hence, find the area of the triangle NPT.

Example 11

The point $P(at^2, 2at)$, lies on the parabola C with equation $y^2 = 4ax$ where a is a positive constant. Show that an equation of the normal to C at P is $y + tx = 2at + at^3$.

C: $y^2 = 4ax$

$y = \sqrt{4ax} = 2\sqrt{a}\sqrt{x} = 2\sqrt{a}\,x^{\frac{1}{2}}$

Rearrange the equation for C in the form $y = kx^n$.
$\sqrt{4ax} = \sqrt{2} \times \sqrt{a} \times \sqrt{x}$

So $y = 2\sqrt{a}\,x^{\frac{1}{2}}$

$\frac{dy}{dx} = 2\sqrt{a}\left(\frac{1}{2}\right)x^{-\frac{1}{2}} = \frac{2\sqrt{a}}{2}x^{-\frac{1}{2}}$

Differentiate to determine the gradient of C.

So, $\frac{dy}{dx} = \frac{\sqrt{a}}{\sqrt{x}}$

Simplify $\frac{dy}{dx}$

At P, $x = at^2$ and

$m_T = \frac{dy}{dx} = \frac{\sqrt{a}}{\sqrt{at^2}} = \frac{\sqrt{a}}{\sqrt{a}t} = \frac{1}{t}$

Substitute $x = at^2$, to calculate the gradient of the tangent to C.

Gradient of tangent at P is $m_T = \frac{1}{t}$.

So gradient of normal is $m_N = -t$.

Use $m_N = \frac{-1}{\left(\frac{1}{t}\right)} = -1 \times \frac{t}{1} = -t$.

P has coordinates $(at^2, 2at)$.

N: $y - 2at = -t(x - at^2)$.

Use $y - y_1 = m_N(x - x_1)$ to find the equation of the normal, **N**. Here $m_N = -t$ and $(x_1, y_1) = (at^2, 2at)$.

N: $y - 2at = -tx + at^3$

Multiply out brackets.

N: $y + tx = 2at + at^3$

Rearrange into the required form.

Therefore, the equation of the normal to C at P is
$y + tx = 2at + at^3$

Example 12

The point $P\left(ct, \frac{c}{t}\right)$, $t \neq 0$, lies on the rectangular hyperbola H with equation $xy = c^2$ where c is a positive constant.

a Show that an equation of the tangent to H at P is $x + t^2y = 2ct$.

A rectangular hyperbola G has equation $xy = 9$. The tangent to G at the point A and the tangent to G at the point B meet at the point $(-1, 7)$.

b Find the coordinates of A and B.

a H: $xy = c^2$

$$y = \frac{c^2}{x} \Rightarrow y = c^2x^{-1}$$

Rearrange the equation for H in the form $y = x^n$.

$$\frac{dy}{dx} = -c^2x^{-2} = -\frac{c^2}{x^2}$$

Differentiate to determine the gradient of H.

At P, $x = ct$ and

$$m_T = \frac{dy}{dx} = -\frac{c^2}{(ct)^2} = -\frac{c^2}{c^2t^2} = \frac{-1}{t^2}$$

Substitute $x = ct$, to calculate the gradient of the tangent to H.

Gradient of tangent at P is

$$m_T = \frac{-1}{t^2}.$$

P has coordinates $\left(ct, \frac{c}{t}\right)$.

T: $y - \frac{c}{t} = -\frac{1}{t^2}(x - ct)$

Use $y - y_1 = m_T(x - x_1)$ to find the equation of the tangent, **T**. Here $m_T = -\frac{1}{t^2}$ and $(x_1, y_1) = \left(ct, \frac{c}{t}\right)$.

T: $t^2y - ct = -(x - ct)$

Multiply both sides by t^2.

T: $t^2y - ct = -x + ct$

Multiply out brackets.

T: $x + t^2y = 2ct$

Rearrange into the required form.

Therefore, the equation of the tangent to H at P is $x + t^2y = 2ct$.

b Compare G: $xy = 9$ with $xy = c^2$

So, as c is positive, $c = 3$.

$c^2 = 9 \Rightarrow c = \sqrt{9} \Rightarrow c = 3$.

Tangent to G is $x + t^2y = 6t$ (1)

Substitute $c = 3$ into the equation of the tangent derived in **a**.

$$-1 + t^2(7) = 6t$$

Substitute $x = -1$ and $y = 7$ in (**1**) as tangent goes through point $(-1, 7)$.

$$7t^2 - 6t - 1 = 0$$

Rearrange into a 'quadratic equation = 0.'

$$(7t + 1)(t - 1) = 0$$

Factorise.

$$t = -\tfrac{1}{7}, 1$$

Solve.

P has coordinates

$$\left(ct, \frac{c}{t}\right) = \left(3t, \frac{3}{t}\right).$$

Substitute $c = 3$ into the general coordinates of P.

When $t = -\frac{1}{7}$, the coordinates are

$$\left(3\left(-\frac{1}{7}\right), \frac{3}{-\frac{1}{7}}\right) = \left(-\frac{3}{7}, -21\right).$$

Substitute $t = -\frac{1}{7}$ into P.

When $t = 1$, the coordinates are

$$\left(3(1), \frac{3}{(1)}\right) = (3, 3).$$

Substitute $t = 1$ into P.

Therefore, the coordinates of A and B are $\left(-\frac{3}{7}, -21\right)$ and $(3, 3)$.

Exercise 3E

1 The point $P(3t^2, 6t)$ lies on the parabola C with equation $y^2 = 12x$.

a Show that an equation of the tangent to C at P is $yt = x + 3t^2$.

b Show that an equation of the normal to C at P is $xt + y = 3t^3 + 6t$.

2 The point $P\left(6t, \frac{6}{t}\right)$, $t \neq 0$, lies on the rectangular hyperbola H with equation $xy = 36$.

a Show that an equation of the tangent to H at P is $x + t^2y = 12t$.

b Show that an equation of the normal to H at P is $t^3x - ty = 6(t^4 - 1)$.

3 The point $P(5t^2, 10t)$ lies on the parabola C with equation $y^2 = 4ax$, where a is a constant and $t \neq 0$.

a Find the value of a.

b Show that an equation of the tangent to C at P is $yt = x + 5t^2$.

The tangent to C at P cuts the x-axis at the point X and the y-axis at the point Y. The point O is the origin of the coordinate system.

c Find, in terms of t, the area of the triangle OXY.

4 The point $P(at^2, 2at)$, $t \neq 0$, lies on the parabola C with equation $y^2 = 4ax$, where a is a positive constant.

a Show that an equation of the tangent to C at P is $ty = x + at^2$.

The tangent to C at the point A and the tangent to C at the point B meet at the point with coordinates $(-4a, 3a)$.

b Find, in terms of a, the coordinates of A and the coordinates of B.

5 The point $P\left(4t, \frac{4}{t}\right)$, $t \neq 0$, lies on the rectangular hyperbola H with equation $xy = 16$.

a Show that an equation of the tangent to C at P is $x + t^2y = 8t$.

The tangent to H at the point A and the tangent to H at the point B meet at the point X with y-coordinate 5. X lies on the directrix of the parabola C with equation $y^2 = 16x$.

b Write down the coordinates of X.

c Find the coordinates of A and B.

d Deduce the equations of the tangents to H which pass through X. Give your answers in the form $ax + by + c = 0$, where a, b and c are integers.

6 The point $P(at^2, 2at)$ lies on the parabola C with equation $y^2 = 4ax$, where a is a constant and $t \neq 0$. The tangent to C at P cuts the x-axis at the point A.

a Find, in terms of a and t, the coordinates of A.

The normal to C at P cuts the x-axis at the point B.

b Find, in terms of a and t, the coordinates of B.

c Hence find, in terms of a and t, the area of the triangle APB.

7 The point $P(2t^2, 4t)$ lies on the parabola C with equation $y^2 = 8x$.

a Show that an equation of the normal to C at P is $xt + y = 2t^3 + 4t$.

The normals to C at the points R, S and T meet at the point $(12, 0)$.

b Find the coordinates of R, S and T.

c Deduce the equations of the normals to C which all pass through the point $(12, 0)$.

8 The point $P(at^2, 2at)$ lies on the parabola C with equation $y^2 = 4ax$, where a is a positive constant and $t \neq 0$. The tangent to C at P meets the y-axis at Q.

a Find in terms of a and t, the coordinates of Q.

The point S is the focus of the parabola.

b State the coordinates of S.

c Show that PQ is perpendicular to SQ.

9 The point $P(6t^2, 12t)$ lies on the parabola C with equation $y^2 = 24x$.

a Show that an equation of the tangent to the parabola at P is $ty = x + 6t^2$.

The point X has y-coordinate 9 and lies on the directrix of C.

b State the x-coordinate of X.

The tangent at the point B on C goes through point X.

c Find the possible coordinates of B.

Mixed exercise 3F

1 A parabola C has equation $y^2 = 12x$. The point S is the focus of C.

a Find the coordinates of S.

The line l with equation $y = 3x$ intersects C at the point P where $y > 0$.

b Find the coordinates of P.

c Find the area of the triangle OPS, where O is the origin.

2 A parabola C has equation $y^2 = 24x$. The point P with coordinates $(k, 6)$, where k is a constant lies on C.

a Find the value of k.

The point S is the focus of C.

b Find the coordinates of S.

The line l passes through S and P and intersects the directrix of C at the point D.

c Show that an equation for l is $4x + 3y - 24 = 0$.

d Find the area of the triangle OPD, where O is the origin.

3 The parabola C has parametric equations $x = 12t^2$, $y = 24t$. The focus to C is at the point S.

a Find a Cartesian equation of C.

The point P lies on C where $y > 0$. P is 28 units from S.

b Find an equation of the directrix of C.

c Find the exact coordinates of the point P.

d Find the area of the triangle OSP, giving your answer in the form $k\sqrt{3}$, where k is an integer.

4 The point $(4t^2, 8t)$ lies on the parabola C with equation $y^2 = 16x$. The line l with equation $4x - 9y + 32 = 0$ intersects the curve at the points P and Q.

a Find the coordinates of P and Q.

b Show that an equation of the normal to C at $(4t^2, 8t)$ is $xt + y = 4t^3 + 8t$.

c Hence, find an equation of the normal to C at P and an equation of the normal to C at Q.

The normal to C at P and the normal to C at Q meet at the point R.

d Find the coordinates of R and show that R lies on C.

e Find the distance OR, giving your answer in the form $k\sqrt{97}$, where k is an integer.

5 The point $P\ (at^2, 2at)$ lies on the parabola C with equation $y^2 = 4ax$, where a is a positive constant. The point Q lies on the directrix of C. The point Q also lies on the x-axis.

a State the coordinates of the focus of C and the coordinates of Q.

The tangent to C at P passes through the point Q.

b Find, in terms of a, the two sets of possible coordinates of P.

6 The point $P\left(ct, \frac{c}{t}\right)$, $c > 0$, $t \neq 0$, lies on the rectangular hyperbola H with equation $xy = c^2$.

a Show that the equation of the normal to H at P is $t^3x - ty = c(t^4 - 1)$.

b Hence, find the equation of the normal n to the curve V with the equation $xy = 36$ at the point $(12, 3)$. Give your answer in the form $ax + by = d$, where a, b and d are integers.

The line n meets V again at the point Q.

c Find the coordinates of Q.

7 A rectangular hyperbola H has equation $xy = 9$. The lines l_1 and l_2 are tangents to H. The gradients of l_1 and l_2 are both $-\frac{1}{4}$. Find the equations of l_1 and l_2.

8 The point P lies on the rectangular hyperbola $xy = c^2$, where $c > 0$. The tangent to the rectangular hyperbola at the point $P\left(ct, \frac{c}{t}\right)$, $t > 0$, cuts the x-axis at the point X and cuts the y-axis at the point Y.

a Find, in terms of c and t, the coordinates of X and Y.

b Given that the area of the triangle OXY is 144, find the exact value of c.

9 The points $P(4at^2, 4at)$ and $Q(16at^2, 8at)$ lie on the parabola C with equation $y^2 = 4ax$, where a is a positive constant.

a Show that an equation of the tangent to C at P is $2ty = x + 4at^2$.

b Hence, write down the equation of the tangent to C at Q.

The tangent to C at P meets the tangent to C at Q at the point R.

c Find, in terms of a and t, the coordinates of R.

10 A rectangular hyperbola H has Cartesian equation $xy = c^2$, $c > 0$. The point $\left(ct, \frac{c}{t}\right)$, where $t \neq 0$, $t > 0$ is a general point on H.

a Show that an equation an equation of the tangent to H at $\left(ct, \frac{c}{t}\right)$ is $x + t^2y = 2ct$.

The point P lies on H. The tangent to H at P cuts the x-axis at the point X with coordinates $(2a, 0)$, where a is a constant.

b Use the answer to part **a** to show that P has coordinates $\left(a, \frac{c^2}{a}\right)$.

The point Q, which lies on H, has x-coordinate $2a$.

c Find the y-coordinate of Q.

d Hence, find the equation of the line OQ, where O is the origin.

The lines OQ and XP meet at point R.

e Find, in terms of a, the x-coordinate of R.

Given that the line OQ is perpendicular to the line XP,

f Show that $c^2 = 2a^2$,

g find, in terms of a, the y-coordinate of R.

Summary of key points

1 To find the Cartesian equation of a curve given parametrically you eliminate the parameter t between the parametric equations.

2 A parabola is a set of points which are equidistant from the focus S and a line called the directrix.

So, for the **parabola** opposite,

- $SP = PX$.
- the **focus**, S, has coordinates $(a, 0)$.
- the **directrix** has equation $x + a = 0$.

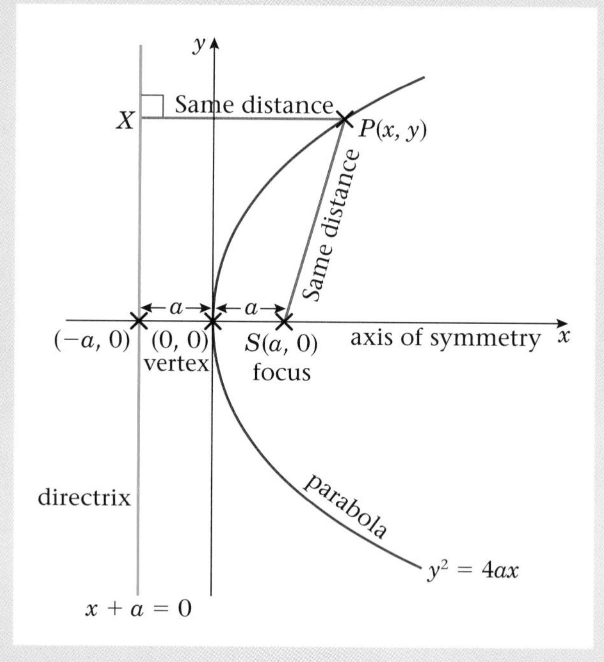

3 The curve opposite is a sketch of a **parabola** with a Cartesian equation of $y^2 = 4ax$, where a is a positive constant.

This curve has parametric equations:

$$x = at^2, y = 2at, t \in \mathbb{R}.$$

where a is a positive constant.

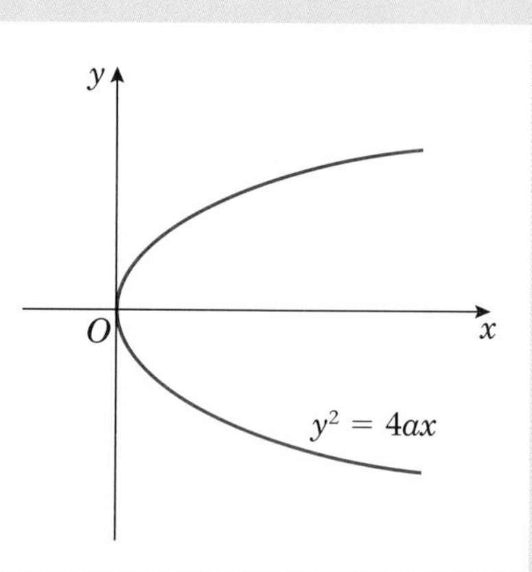

4 The curve opposite is a sketch of a **rectangular hyperbola** with a Cartesian equation of $xy = c^2$, where c is a positive constant.

This curve has parametric equations:

$$x = ct, y = \frac{c}{t}, t \in \mathbb{R}, t \neq 0,$$

where c is a positive constant.

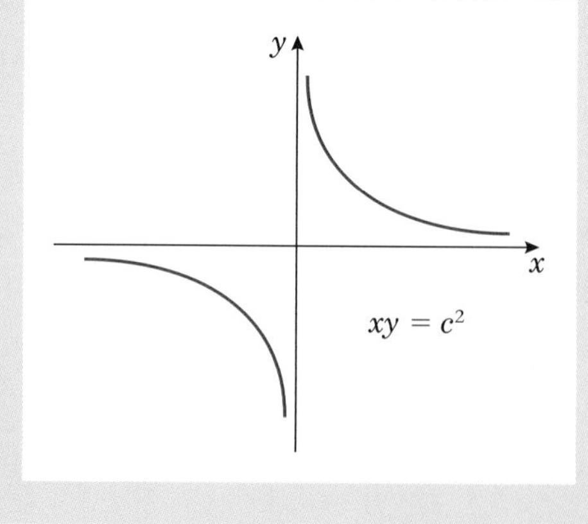

Review Exercise

1 $z_1 = 2 + i$, $z_2 = 3 + 4i$. Find the modulus and the tangent of the argument of each of

a $z_1 z_2^*$ **b** $\frac{z_1}{z_2}$ *E*

2 **a** Show that the complex number $\frac{2 + 3i}{5 + i}$ can be expressed in the form $\lambda(1 + i)$, stating the value of λ.

b Hence show that $\left(\frac{2 + 3i}{5 + i}\right)^4$ is real and determine its value. *E*

3 $z_1 = 5 + i$, $z_2 = -2 + 3i$

a Show that $|z_1|^2 = 2|z_2|^2$.

b Find $\arg(z_1 z_2)$. *E*

4 **a** Find, in the form $p + iq$ where p and q are real, the complex number z which satisfies the equation $\frac{3z - 1}{2 - i} = \frac{4}{1 + 2i}$.

b Show on a single Argand diagram the points which represent z and z^*.

c Express z and z^* in modulus–argument form, giving the arguments to the nearest degree. *E*

5 $z_1 = -1 + i\sqrt{3}$, $z_2 = \sqrt{3} + i$

a Find **i** $\arg z_1$ **ii** $\arg z_2$.

b Express $\frac{z_1}{z_2}$ in the form $a + ib$, where a and b are real, and hence find $\arg\left(\frac{z_1}{z_2}\right)$.

c Verify that $\arg\left(\frac{z_1}{z_2}\right) = \arg z_1 - \arg z_2$. *E*

6 **a** Find the two square roots of $3 - 4i$ in the form $a + ib$, where a and b are real.

b Show the points representing the two square roots of $3 - 4i$ in a single Argand diagram. *E*

7 The complex number z is $-9 + 17i$.

a Show z on an Argand diagram.

b Calculate $\arg z$, giving your answer in radians to two decimal places.

c Find the complex number w for which $zw = 25 + 35i$, giving your answer in the form $p + iq$, where p and q are real. *E*

8 The complex numbers z and w satisfy the simultaneous equations

$2z + \mathrm{i}w = -1, \; z - w = 3 + 3\mathrm{i}$.

a Use algebra to find z, giving your answer in the form $a + \mathrm{i}b$, where a and b are real.

b Calculate $\arg z$, giving your answer in radians to two decimal places.

9 The complex number z satisfies the equation $\dfrac{z - 2}{z + 3\mathrm{i}} = \lambda\mathrm{i}, \; \lambda \in \mathbb{R}$.

a Show that $z = \dfrac{(2 - 3\lambda)(1 + \lambda\mathrm{i})}{1 + \lambda^2}$.

b In the case when $\lambda = 1$, find $|z|$ and $\arg z$.

10 The complex number z is given by $z = -2 + 2\mathrm{i}$.

a Find the modulus and argument of z.

b Find the modulus and argument of $\dfrac{1}{z}$.

c Show on an Argand diagram the points A, B and C representing the complex numbers z, $\dfrac{1}{z}$ and $z + \dfrac{1}{z}$ respectively.

d State the value of $\angle ACB$.

11 The complex numbers z_1 and z_2 are given by $z_1 = \sqrt{3} + \mathrm{i}$ and $z_2 = 1 - \mathrm{i}$.

a Show, on an Argand diagram, points representing the complex numbers z_1, z_2 and $z_1 + z_2$.

b Express $\dfrac{1}{z_1}$ and $\dfrac{1}{z_2}$, each in the form $a + \mathrm{i}b$, where a and b are real numbers.

c Find the values of the real numbers A and B such that $\dfrac{A}{z_1} + \dfrac{B}{z_2} = z_1 + z_2$.

12 The complex numbers z and w are given by $z = \dfrac{A}{1 - \mathrm{i}}$, $w = \dfrac{B}{1 - 3\mathrm{i}}$, where A and B are real numbers. Given that $z + w = \mathrm{i}$,

a find the value of A and the value of B.

b For these values of A and B, find $\tan[\arg(w - z)]$.

13 **a** Given that $z = 2 - \mathrm{i}$, show that $z^2 = 3 - 4\mathrm{i}$.

b Hence, or otherwise, find the roots, z_1 and z_2, of the equation $(z + \mathrm{i})^2 = 3 - 4\mathrm{i}$.

c Show points representing z_1 and z_2 on a single Argand diagram.

d Deduce that $|z_1 - z_2| = 2\sqrt{5}$.

e Find the value of $\arg(z_1 + z_2)$.

14 **a** Find the roots of the equation $z^2 + 4z + 7 = 0$, giving your answers in the form $p + \mathrm{i}\sqrt{q}$, where p and q are integers.

b Show these roots on an Argand diagram.

c Find for each root,

i the modulus,

ii the argument, in radians, giving your answers to three significant figures.

15 Given that $\lambda \in \mathbb{R}$ and that z and w are complex numbers, solve the simultaneous equations $z - \mathrm{i}w = 2, \; z - \lambda w = 1 - \lambda^2$, giving your answers in the form $a + \mathrm{i}b$, where $a, b \in \mathbb{R}$, and a and b are functions of λ.

16 Given that $z_1 = 5 - 2\mathrm{i}$,

a evaluate $|z_1|$, giving your answer as a surd,

b find, in radians to two decimal places, $\arg z_1$.

Given also that z_1 is a root of the equation $z^2 - 10z + c = 0$, where c is a real number,

c find the value of c.

17 The complex numbers z and w are given by $z = \frac{5 - 10i}{2 - i}$ and $w = iz$.

a Obtain z and w in the form $p + iq$, where p and q are real numbers.

b Show points representing z and w on a single Argand diagram

The origin O and the points representing z and w are the vertices of a triangle.

c Show that this triangle is isosceles and state the angle between the equal sides. **E**

18 $z_1 = \frac{1 + i}{1 - i}$, $z_2 = \frac{\sqrt{2}}{1 - i}$

a Find the modulus and argument of each of the complex numbers z_1 and z_2.

b Plot the points representing z_1, z_2 and $z_1 + z_2$ on a single Argand diagram.

c Deduce from your diagram that $\tan\left(\frac{3\pi}{8}\right) = 1 + \sqrt{2}$. **E**

19 $z_1 = 1 + 2i$, $z_2 = \frac{3}{5} + \frac{4}{5}i$

a Express in the form $p + qi$, where $p, q \in \mathbb{R}$,

i z_1z_2 **ii** $\frac{z_1}{z_2}$.

In an Argand diagram, the origin O and the points representing z_1z_2, $\frac{z_1}{z_2}$ and z_3 are the vertices of a rhombus.

b Sketch the rhombus on an Argand diagram.

c Find z_3.

d Show that $|z_3| = \frac{6\sqrt{5}}{5}$. **E**

20 $z_1 = -30 + 15i$.

a Find $\arg z_1$, giving your answer in radians to two decimal places.

The complex numbers z_2 and z_3 are given by $z_2 = -3 + pi$ and $z_3 = q + 3i$, where p and q are real constants and $p > q$.

b Given that $z_2 z_3 = z_1$, find the value of p and the value of q.

c Using your values of p and q, plot the points corresponding to z_1, z_2 and z_3 on an Argand diagram.

d Verify that $2z_2 + z_3 - z_1$ is real and find its value. **E**

21 Given that $z = 1 + \sqrt{3}i$ and that $\frac{w}{z} = 2 + 2i$, find

a w in the form $a + ib$, where $a, b \in \mathbb{R}$,

b the argument of w,

c the exact value for the modulus of w.

On an Argand diagram, the point A represents z and the point B represents w.

d Draw the Argand diagram, showing the points A and B.

e Find the distance AB, giving your answer as a simplified surd. **E**

22 The solutions of the equation $z^2 + 6z + 25 = 0$ are z_1 and z_2, where $0 < \arg z_1 < \pi$ and $-\pi < \arg z_2 < 0$.

a Express z_1 and z_2 in the form $a + ib$, where a and b are integers.

b Show that $z_1^2 = -7 - 24i$.

c Find $|z_1^2|$.

d Find $\arg\left(z_1^2\right)$.

e Show, on an Argand diagram, the points which represent the complex numbers z_1, z_2 and z_1^2. **E**

23 $z = \sqrt{3} - i$. z^* is the complex conjugate of z.

a Show that $\frac{z}{z^*} = \frac{1}{2} - \frac{\sqrt{3}}{2}i$.

b Find the value of $\left|\frac{z}{z^*}\right|$.

c Verify, for $z = \sqrt{3} - i$, that $\arg \frac{z}{z^*} = \arg z - \arg z^*$.

d Display z, z^* and $\frac{z}{z^*}$ on a single Argand diagram.

e Find a quadratic equation with roots z and z^* in the form $ax^2 + bx + c = 0$, where a, b and c are real constants to be found. **E**

24 $z = \frac{1 + 7i}{4 + 3i}$.

a Find the modulus and argument of z.

b Write down the modulus and argument of z^*.

In an Argand diagram, the points A and B represent $1 + 7i$ and $4 + 3i$ respectively and O is the origin. The quadrilateral $OABC$ is a parallelogram.

c Find the complex number represented by the point C.

d Calculate the area of the parallelogram. **E**

25 Given that $\frac{z + 2i}{z - \lambda i} = i$, where λ is a positive, real constant,

a show that $z = \left(\frac{\lambda}{2} + 1\right) + i\left(\frac{\lambda}{2} - 1\right)$.

Given also that $\tan(\arg z) = \frac{1}{2}$, calculate

b the value of λ,

c the value of $|z|^2$. **E**

26 The complex numbers $z_1 = 2 + 2i$ and $z_2 = 1 + 3i$ are represented on an Argand diagram by the points P and Q respectively.

a Display z_1 and z_2 on the same Argand diagram.

b Find the exact values of $|z_1|$, $|z_2|$ and the length of PQ.

Hence show that

c $\triangle OPQ$, where O is the origin, is right-angled.

Given that $OPQR$ is a rectangle in the Argand diagram,

d find the complex number z_3 represented by the point R. **E**

27 The complex number z is given by $z = (1 + 3i)(p + qi)$, where p and q are real and $p > 0$.

Given that $\arg z = \frac{\pi}{4}$,

a show that $p + 2q = 0$.

Given also that $|z| = 10\sqrt{2}$,

b find the value of p and the value of q.

c Write down the value of $\arg z^*$. **E**

28 The complex numbers z_1 and z_2 are given by $z_1 = 5 + i$, $z_2 = 2 - 3i$.

a Show points representing z_1 and z_2 on an Argand diagram.

b Find the modulus of $z_1 - z_2$.

c Find the complex number $\frac{z_1}{z_2}$ in the form $a + ib$, where a and b are rational numbers.

d Hence find the argument of $\frac{z_1}{z_2}$, giving your answer in radians to three significant figures.

e Determine the values of the real constants p and q such that

$\frac{p + iq + 3z_1}{p - iq + 3z_2} = 2i$. **E**

29 $z = a + ib$, where a and b are real and non-zero.

a Find z^2 and $\frac{1}{z}$ in terms of a and b, giving each answer in the form $x + iy$, where x and y are real.

b Show that $|z^2| = a^2 + b^2$.

c Find $\tan(\arg z^2)$ and $\tan\left(\arg \frac{1}{z}\right)$, in terms of a and b.

On an Argand diagram the point P represents z^2 and the point Q represents $\frac{1}{z}$ and O the origin.

d Using your answer to **c**, or otherwise, show that if P, O and Q are collinear, then $3a^2 = b^2$. **E**

30 Starting with $x = 1.5$, apply the Newton–Raphson procedure once to $f(x) = x^3 - 3$ to obtain a better approximation to the cube root of 3, giving your answer to three decimal places. **E**

31 $f(x) = 2^x + x - 4$. The equation $f(x) = 0$ has a root α in the interval [1, 2]. Use linear interpolation on the values at the end points of this interval to find an approximation to α. **E**

32 Given that the equation $x^3 - x - 1 = 0$ has a root near 1.3, apply the Newton–Raphson procedure once to $f(x) = x^3 - x - 1$ to obtain a better approximation to this root, giving your answer to three decimal places. **E**

33 $f(x) = x^3 - 12x + 7$.

a Use differentiation to find $f'(x)$.

The equation $f(x) = 0$ has a root α in the interval $\frac{1}{2} < x < 1$.

b Taking $x = \frac{1}{2}$ as a first approximation to α, use the Newton–Raphson procedure twice to obtain two further approximations to α. Give your final answer to four decimal places. **E**

34 The equation $\sin x = \frac{1}{2}x$ has a root in the interval [1.8, 2]. Use linear interpolation once on the interval [1.8, 2] to find an estimate of the root, giving your answer to two decimal places.

35 $f(x) = x^4 + 3x^3 - 4x - 5$. The equation $f(x) = 0$ has a root between $x = 1.2$ and $x = 1.6$. Starting with the interval [1.2, 1.6], use interval bisection three times to obtain an interval of width 0.05 which contains this root.

36 $f(x) = 3\tan\left(\frac{x}{2}\right) - x - 1, -\pi < x < \pi$.

Given that $f(x) = 0$ has a root between 1 and 2, use linear interpolation once on the interval [1, 2] to find an approximation to this root. Give your answer to two decimal places. **E**

37 $f(x) = 3^x - x - 6$.

a Show that $f(x) = 0$ has a root α between $x = 1$ and $x = 2$.

b Starting with the interval [1, 2], use interval bisection three times to find an interval of width 0.125 which contains α. **E**

38 Given that x is measured in radians and $f(x) = \sin x - 0.4x$,

a find the values of f(2) and f(2.5) and deduce that the equation $f(x) = 0$ has a root α in the interval [2, 2.5],

b use linear interpolation once on the interval [2, 2.5] to estimate the value of α, giving your answer to two decimal places. **E**

39 $f(x) = \tan x + 1 - 4x^2, -\frac{\pi}{2} < x < \frac{\pi}{2}$.

a Show that $f(x) = 0$ has a root α in the interval [1.42, 1.44].

b Use linear interpolation once on the interval [1.42, 1.44] to find an estimate of α, giving your answer to three decimal places. **E**

40 $f(x) = \cos\sqrt{x} - x$

a Show that $f(x) = 0$ has a root α in the interval [0.5, 1].

b Use linear interpolation on the interval [0.5, 1] to obtain an approximation to α. Give your answer to two decimal places.

c By considering the change of sign of $f(x)$ over an appropriate interval, show that your answer to **b** is accurate to two decimal places.

41 $f(x) = 2^x - x^2 - 1$
The equation $f(x) = 0$ has a root α between $x = 4.256$ and $x = 4.26$.

a Starting with the interval [4.256, 4.26] use interval bisection three times to find an interval of width 5×10^{-4} which contains α.

b Write down the value of α, correct to three decimal places.

42 $f(x) = 2x^2 + \frac{1}{x} - 3$
The equation $f(x) = 0$ has a root α in the interval $0.3 < x < 0.5$.

a Use linear interpolation once on the interval $0.3 < x < 0.5$ to find an approximation to α. Give your answer to three decimal places.

b Find $f'(x)$.

c Taking 0.4 as an approximation to α, use the Newton–Raphson procedure once to find another approximation to α.

43 $f(x) = 0.25x - 2 + 4 \sin \sqrt{x}$.

a Show that the equation $f(x) = 0$ has a root α between $x = 0.24$ and $x = 0.28$.

b Starting with the interval [0.24, 0.28], use interval bisection three times to find an interval of width 0.005 which contains α **E**

44 $f(x) = x^3 + 8x - 19$.

a Show that the equation $f(x) = 0$ has only one real root.

b Show that the real root of $f(x) = 0$ lies between 1 and 2.

c Obtain an approximation to the real root of $f(x) = 0$ by performing two applications of the Newton–Raphson procedure to $f(x)$, using $x = 2$ as the first approximation. Give your answer to three decimal places.

d By considering the change of sign of $f(x)$ over an appropriate interval, show that your answer to **c** is accurate to three decimal places. **E**

45 $f(x) = x^3 - 3x - 1$
The equation $f(x) = 0$ has a root α in the interval $[-2, -1]$.

a Use linear interpolation on the values at the ends of the interval $[-2, -1]$ to obtain an approximation to α.

The equation $f(x) = 0$ has a root β in the interval $[-1, 0]$.

b Taking $x = -0.5$ as a first approximation to β, use the Newton–Raphson procedure once to obtain a second approximation to β.

The equation $f(x) = 0$ has a root γ in the interval [1.8, 1.9].

c Starting with the interval [1.8, 1.9] use interval bisection twice to find an interval of width 0.025 which contains γ.

46 A point P with coordinates (x, y) moves so that its distance from the point (5, 0) is equal to its distance from the line with equation $x = -5$.

Prove that the locus of P has an equation of the form $y^2 = 4ax$, stating the value of a.

47 A parabola C has equation $y^2 = 16x$. The point S is the focus of the parabola.

a Write down the coordinates of S.

The point P with coordinates (16, 16) lies on C.

b Find an equation of the line SP, giving your answer in the form $ax + by + c = 0$, where a, b and c are integers.

The line SP intersects C at the point Q, where P and Q are distinct points.

c Find the coordinates of Q.

48 The curve C has equations $x = 3t^2$, $y = 6t$.

a Sketch the graph of the curve C.

The curve C intersects the line with equation $y = x - 72$ at the points A and B.

b Find the length AB, giving your answer as a surd in its simplest form.

49 A parabola C has equation $y^2 = 12x$. The points P and Q both lie on the parabola and are both at a distance 8 from the directrix of the parabola. Find the length PQ, giving your answer in surd form.

50 The point $P(2, 8)$ lies on the parabola C with equation $y^2 = 4ax$. Find

a the value of a,

b an equation of the tangent to C at P.

The tangent to C at P cuts the x-axis at the point X and the y-axis at the point Y.

c Find the exact area of the triangle OXY.

51 The point P with coordinates $(3, 4)$ lies on the rectangular hyperbola H with equation $xy = 12$. The point Q has coordinates $(-2, 0)$. The points P and Q lie on the line l.

a Find an equation of l, giving your answer in the form $y = mx + c$, where m and c are real constants.

The line l cuts H at the point R, where P and R are distinct points.

b Find the coordinates of R.

52 The point $P(12, 3)$ lies on the rectangular hyperbola H with equation $xy = 36$.

a Find an equation of the tangent to H at P.

The tangent to H at P cuts the x-axis at the point M and the y-axis at the point N.

b Find the length MN, giving your answer as a simplified surd.

53 The point $P(5, 4)$ lies on the rectangular hyperbola H with equation $xy = 20$. The line l is the normal to H at P.

a Find an equation of l, giving your answer in the form $ax + by + c = 0$, where a, b and c are integers.

The line l meets H again at the point Q.

b Find the coordinates of Q.

54 The curve H with equation $x = 8t$, $y = \frac{16}{t}$ intersects the line with equation $y = \frac{1}{4}x + 4$ at the points A and B. The mid-point of AB is M. Find the coordinates of M.

55 The point $P(24t^2, 48t)$ lies on the parabola with equation $y^2 = 96x$. The point P also lies on the rectangular hyperbola with equation $xy = 144$.

a Find the value of t and, hence, the coordinates of P.

b Find an equation of the tangent to the parabola at P, giving your answer in the form $y = mx + c$, where m and c are real constants.

c Find an equation of the tangent to the rectangular hyperbola at P, giving your answer in the form $y = mx + c$, where m and c are real constants.

56 The points $P(9, 8)$ and $Q(6, 12)$ lie on the rectangular hyperbola H with equation $xy = 72$.

a Show that an equation of the chord PQ of H is $4x + 3y = 60$.

The point R lies on H. The tangent to H at R is parallel to the chord PQ.

b Find the exact coordinates of the two possible positions of R.

57 A rectangular hyperbola H has cartesian equation $xy = 9$. The point $\left(3t, \frac{3}{t}\right)$ is a general point on H.

a Show that an equation of the tangent to H at $\left(3t, \frac{3}{t}\right)$ is $x + t^2y = 6t$.

The tangent to H at $\left(3t, \frac{3}{t}\right)$ cuts the x-axis at A and the y-axis at B. The point O is the origin of the coordinate system.

b Show that, as t varies, the area of the triangle OAB is constant. **E**

58 The point $P\left(ct, \frac{c}{t}\right)$ lies on the hyperbola with equation $xy = c^2$, where c is a positive constant.

a Show that an equation of the normal to the hyperbola at P is
$t^3x - ty - c(t^4 - 1) = 0$.

The normal to the hyperbola at P meets the line $y = x$ at G. Given that $t \neq \pm 1$,

b show that $PG^2 = c^2\left(t^2 + \frac{1}{t^2}\right)$. **E**

59 **a** Show that an equation of the tangent to the rectangular hyperbola with equation $xy = c^2$ at the point $\left(ct, \frac{c}{t}\right)$ is $t^2y + x = 2ct$.

Tangents are drawn from the point $(-3, 3)$ to the rectangular hyperbola with equation $xy = 16$.

b Find the coordinates of the points of contact of these tangents with the hyperbola. **E**

60 The point $P\ (at^2, 2at)$, where $t > 0$, lies on the parabola with equation $y^2 = 4ax$. The tangent and normal at P cut the x-axis at the points T and N respectively. Prove that $\frac{PT}{PN} = t$. **E**

61 The point P lies on the parabola with equation $y^2 = 4ax$, where a is a positive constant.

a Show that an equation of the tangent to the parabola $P\ (ap^2, 2ap)$, $p > 0$, is $py = x + ap^2$.

The tangents at the points $P\ (ap^2, 2ap)$ and $Q\ (aq^2, 2aq)$ $(p \neq q, p > 0, q > 0)$ meet at the point N.

b Find the coordinates of N.

Given further that N lies on the line with equation $y = 4a$,

c find p in terms of q. **E**

62 The point $P\ (at^2, 2at)$, $t \neq 0$ lies on the parabola with equation $y^2 = 4ax$, where a is a positive constant.

a Show that an equation of the normal to the parabola at P is
$y + xt = 2at + at^3$.

The normal to the parabola at P meets the parabola again at Q.

b Find, in terms of t, the coordinates of Q.

63 **a** Show that the normal to the rectangular hyperbola $xy = c^2$, at the point $P\left(ct, \frac{c}{t}\right)$, $t \neq 0$, has equation $y = t^2x + \frac{c}{t} - ct^3$.

The normal to the hyperbola at P meets the hyperbola again at the point Q.

b Find, in terms of t, the coordinates of the point Q.

Given that the mid-point of PQ is (X, Y) and that $t \neq \pm 1$,

c show that $\frac{X}{Y} = -\frac{1}{t^2}$. **E**

64 The rectangular hyperbola C has equation $xy = c^2$, where c is a positive constant.

a Show that the tangent to C at the point $P\left(cp, \frac{c}{p}\right)$ has equation $p^2y = -x + 2cp$.

The point Q has coordinates $Q\left(cq, \frac{c}{q}\right)$, $q \neq p$.

The tangents to C at P and Q meet at N. Given that $p + q \neq 0$,

b show that the y-coordinate of N is $\frac{2c}{p + q}$.

The line joining N to the origin O is perpendicular to the chord PQ.

c Find the numerical value of p^2q^2. **E**

65 The point P lies on the rectangular hyperbola $xy = c^2$, where c is a positive constant.

a Show that an equation of the tangent to the hyperbola at the point $P\left(cp, \frac{c}{p}\right)$, $p > 0$, is $yp^2 + x = 2cp$.

This tangent at P cuts the x-axis at the point S.

b Write down the coordinates of S.

c Find an expression, in terms of p, for the length of PS.

The normal at P cuts the x-axis at the point R. Given that the area of $\triangle RPS$ is $41c^2$,

d find, in terms of c, the coordinates of the point P. **E**

66 The curve C has equation $y^2 = 4ax$, where a is a positive constant.

a Show that an equation of the normal to C at the point $P\ (ap^2, 2ap)$, $(p \neq 0)$ is $y + px = 2ap + ap^3$.

The normal at P meets C again at the point $Q\ (aq^2, 2aq)$.

b Find q in terms of p.

Given that the mid-point of PQ has coordinates $\left(\frac{125}{18}a,\ -3a\right)$,

c use your answer to **b**, or otherwise, to find the value of p. **E**

67 The parabola C has equation $y^2 = 32x$.

a Write down the coordinates of the focus S of C.

b Write down the equation of the directrix of C.

The points $P\ (2, 8)$ and $Q\ (32, -32)$ lie on C.

c Show that the line joining P and Q goes through S.

The tangent to C at P and the tangent to C at Q intersect at the point D.

d Show that D lies on the directrix of C.

After completing this chapter you should be able to

- add, subtract and multiply matrices
- find inverses of 2×2 matrices
- represent some geometrical transformations with 2×2 matrices
- use matrices to solve linear simultaneous equations.

Matrix algebra

This chapter will give a brief introduction to matrices and some of their applications. Matrix algebra is used in many branches of mathematics, especially those areas where large volumes of data are handled as the rules for combining matrices are easily implemented on computers. Transformation matrices are often used to create 3-D computer graphics. You will meet some further properties and applications if you study FP3.

4.1 You can find the dimension of a matrix.

- A **matrix** can be thought of as an array of numbers (a collection of numbers set out in a table) and they come in different shapes and sizes.

- You can describe these different shapes and sizes in terms of the **dimension** of the matrix. This is given by two numbers n and m in the form $n \times m$ (read as n by m), where n is the number of rows (horizontal or across the page) and m is the number of columns (vertical or down the page) in the matrix.

- An $n \times m$ matrix has n rows and m columns.

- Matrices are usually denoted in bold print with a capital letter e.g. **A**, **M** etc.

Example 1

Give the dimensions of the following matrices

a $\begin{pmatrix} 2 & -1 \\ 1 & 3 \end{pmatrix}$, **b** $(1 \quad 0 \quad 2)$,

c $\begin{pmatrix} 4 \\ -1 \end{pmatrix}$, **d** $\begin{pmatrix} 3 & 2 \\ -1 & 1 \\ 0 & -3 \end{pmatrix}$.

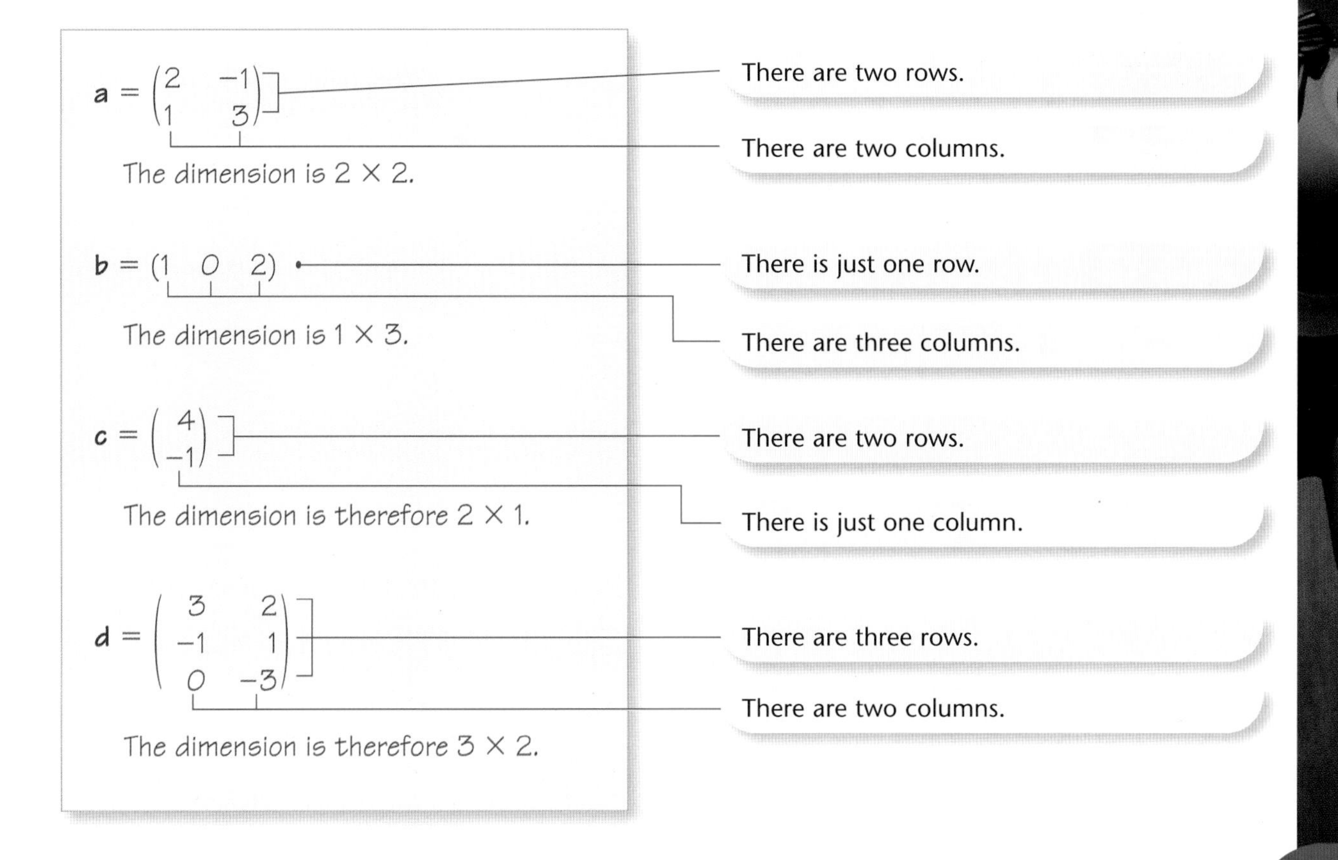

4.2 You can add and subtract matrices of the same dimension.

To **add** or **subtract** matrices you simply add or subtract the corresponding **elements** of the two matrices.

Example 2

Find **a** $\begin{pmatrix} 2 & -1 \\ 0 & 3 \end{pmatrix} + \begin{pmatrix} -1 & 4 \\ 5 & 3 \end{pmatrix}$

b $\begin{pmatrix} 1 & -3 & 4 \\ 2 & 1 & 1 \end{pmatrix} - \begin{pmatrix} 0 & 2 & 1 \\ 5 & 2 & 3 \end{pmatrix}$

a $\begin{pmatrix} 2 & -1 \\ 0 & 3 \end{pmatrix} + \begin{pmatrix} -1 & 4 \\ 5 & 3 \end{pmatrix}$

$= \begin{pmatrix} 1 & 3 \\ 5 & 6 \end{pmatrix}$

$2 + -1 = 1$

$-1 + 4 = 3$

$3 + 3 = 6$

$0 + 5 = 5$

b $\begin{pmatrix} 1 & -3 & 4 \\ 2 & 1 & 1 \end{pmatrix} - \begin{pmatrix} 0 & 2 & 1 \\ 5 & 2 & 3 \end{pmatrix}$

$= \begin{pmatrix} 1 & -5 & 3 \\ -3 & -1 & -2 \end{pmatrix}$

Top row
$1 - 0 = 1$
$-3 - 2 = -5$
$4 - 1 = 3$

Bottom row
$2 - 5 = -3$
$1 - 2 = -1$
$1 - 3 = -2$

Example 3

$\mathbf{A} = \begin{pmatrix} 2 & 3 \\ 1 & a \end{pmatrix}$, $\mathbf{B} = \begin{pmatrix} b & -1 \\ 2 & 4 \end{pmatrix}$, $\mathbf{C} = \begin{pmatrix} 3 & y \\ x & 3 \end{pmatrix}$.

Given that $\mathbf{A} + \mathbf{B} = \mathbf{C}$, find the values of the constants a, b, x and y.

$\mathbf{A} + \mathbf{B} = \mathbf{C} \Rightarrow \begin{pmatrix} 2 & 3 \\ 1 & a \end{pmatrix} + \begin{pmatrix} b & -1 \\ 2 & 4 \end{pmatrix} = \begin{pmatrix} 3 & y \\ x & 3 \end{pmatrix}$

Comparing corresponding elements:

First row, first column:	$2 + b = 3$	so $b = 1$
First row, second column:	$3 + -1 = y$	so $y = 2$
Second row, first column:	$1 + 2 = x$	so $x = 3$
Second row, second column:	$a + 4 = 3$	so $a = -1$

You can describe the position of individual elements in terms of their row and column e.g. second row, first column of **C** is x. By considering the equivalent elements in **A** and **B** you can form the equation for x.

Exercise 4A

1 Describe the dimensions of these matrices.

a $\begin{pmatrix} 1 & 0 \\ -1 & 3 \end{pmatrix}$

b $\begin{pmatrix} 1 \\ 2 \end{pmatrix}$

c $\begin{pmatrix} 1 & 2 & 1 \\ 3 & 0 & -1 \end{pmatrix}$

d $(1 \quad 2 \quad 3)$

e $(3 \quad -1)$

f $\begin{pmatrix} 1 & 0 & 0 \\ 0 & 1 & 0 \\ 0 & 0 & 1 \end{pmatrix}$

2 For the matrices

$\mathbf{A} = \begin{pmatrix} 2 & -1 \\ 1 & 3 \end{pmatrix}$, $\mathbf{B} = \begin{pmatrix} 4 & 1 \\ -1 & -2 \end{pmatrix}$, $\mathbf{C} = \begin{pmatrix} 6 & 0 \\ 0 & 1 \end{pmatrix}$,

find

a $\mathbf{A} + \mathbf{C}$

b $\mathbf{B} - \mathbf{A}$

c $\mathbf{A} + \mathbf{B} - \mathbf{C}$.

3 For the matrices

$\mathbf{A} = \begin{pmatrix} 1 \\ 2 \end{pmatrix}$, $\mathbf{B} = (1 \quad -1)$, $\mathbf{C} = (-1 \quad 1 \quad 0)$,

$\mathbf{D} = (0 \quad 1 \quad -1)$, $\mathbf{E} = \begin{pmatrix} 3 \\ -1 \end{pmatrix}$, $\mathbf{F} = (2 \quad 1 \quad 3)$,

find where possible:

a $\mathbf{A} + \mathbf{B}$

b $\mathbf{A} - \mathbf{E}$

c $\mathbf{F} - \mathbf{D} + \mathbf{C}$

d $\mathbf{B} + \mathbf{C}$

e $\mathbf{F} - (\mathbf{D} + \mathbf{C})$

f $\mathbf{A} - \mathbf{F}$

g $\mathbf{C} - (\mathbf{F} - \mathbf{D})$.

4 Given that $\begin{pmatrix} a & 2 \\ -1 & b \end{pmatrix} - \begin{pmatrix} 1 & c \\ d & -2 \end{pmatrix} = \begin{pmatrix} 5 & 0 \\ 0 & 5 \end{pmatrix}$, find the values of the constants a, b, c and d.

5 Given that $\begin{pmatrix} 1 & 2 & 0 \\ a & b & c \end{pmatrix} + \begin{pmatrix} a & b & c \\ 1 & 2 & 0 \end{pmatrix} = \begin{pmatrix} c & 5 & c \\ c & c & c \end{pmatrix}$, find the values of a, b and c.

6 Given that $\begin{pmatrix} 5 & 3 \\ 0 & -1 \\ 2 & 1 \end{pmatrix} + \begin{pmatrix} a & b \\ c & d \\ e & f \end{pmatrix} = \begin{pmatrix} 7 & 1 \\ 2 & 0 \\ 1 & 4 \end{pmatrix}$, find the values of a, b, c, d, e and f.

4.3 You can multiply a matrix by a scalar (number).

- A **scalar** is something that isn't a matrix.

In FP1 a scalar will simply be a number.

- To multiply a matrix by a number you simply multiply each element of the matrix by that number.

Example 4

$\mathbf{A} = \begin{pmatrix} 1 & 2 \\ -1 & 0 \end{pmatrix}$, $\mathbf{B} = (6 \quad 0 \quad -4)$

Find **a** $2\mathbf{A}$ **b** $\frac{1}{2}\mathbf{B}$

a $2\mathbf{A} = \begin{pmatrix} 2 & 4 \\ -2 & 0 \end{pmatrix}$

Top row
$2 \times 1 = 2$
$2 \times 2 = 4$

Bottom row
$2 \times -1 = -2$
$2 \times 0 = 0$

Note that $2\mathbf{A}$ gives the same answer as $\mathbf{A} + \mathbf{A}$.

b $\frac{1}{2}\mathbf{B} = (3 \quad 0 \quad -2)$

$\frac{1}{2} \times 6 = 3$
$\frac{1}{2} \times 0 = 0$
$\frac{1}{2} \times -4 = -2$

Example 5

$\mathbf{A} = \begin{pmatrix} a & 0 \\ 1 & 2 \end{pmatrix}$, $\mathbf{B} = \begin{pmatrix} 1 & b \\ 0 & 3 \end{pmatrix}$, $\mathbf{C} = \begin{pmatrix} 6 & 6 \\ 1 & c \end{pmatrix}$.

Given that $\mathbf{A} + 2\mathbf{B} = \mathbf{C}$, find the values of the constants a, b and c.

$\mathbf{A} + 2\mathbf{B} = \mathbf{C} \Rightarrow \begin{pmatrix} a & 0 \\ 1 & 2 \end{pmatrix} + 2\begin{pmatrix} 1 & b \\ 0 & 3 \end{pmatrix} = \begin{pmatrix} 6 & 6 \\ 1 & c \end{pmatrix}$

Comparing corresponding elements:

First row, first column: $a + 2 \times 1 = 6 \Rightarrow a = 4$

First row, second column: $0 + 2b = 6 \Rightarrow b = 3$

Second row, second column: $2 + 2 \times 3 = c \Rightarrow c = 8$

This gives four equations. You only need three, so select the relevant ones.

Exercise 4B

1 For the matrices $\mathbf{A} = \begin{pmatrix} 2 & 0 \\ 4 & -6 \end{pmatrix}$, $\mathbf{B} = \begin{pmatrix} 1 \\ -1 \end{pmatrix}$, find

a $3\mathbf{A}$ **b** $\frac{1}{2}\mathbf{A}$ **c** $2\mathbf{B}$.

2 Find the value of k and the value of x so that $\begin{pmatrix} 0 & 1 \\ 2 & 0 \end{pmatrix} + k\begin{pmatrix} 0 & 2 \\ -1 & 0 \end{pmatrix} = \begin{pmatrix} 0 & 7 \\ x & 0 \end{pmatrix}$.

3 Find the values of a, b, c and d so that $2\begin{pmatrix} a & 0 \\ 1 & b \end{pmatrix} - 3\begin{pmatrix} 1 & c \\ d & -1 \end{pmatrix} = \begin{pmatrix} 3 & 3 \\ -4 & -4 \end{pmatrix}$.

4 Find the values of a, b, c and d so that $\begin{pmatrix} 5 & a \\ b & 0 \end{pmatrix} - 2\begin{pmatrix} c & 2 \\ 1 & -1 \end{pmatrix} = \begin{pmatrix} 9 & 1 \\ 3 & d \end{pmatrix}$.

5 Find the value of k so that $\begin{pmatrix} -3 \\ k \end{pmatrix} + k\begin{pmatrix} 2k \\ 2k \end{pmatrix} = \begin{pmatrix} k \\ 6 \end{pmatrix}$.

4.4 You can multiply matrices together.

- **Whilst the rule for adding and subtracting matrices appears to be quite natural, it is the rule for multiplying matrices that gives them their useful properties.**
- **The basic operation consists of multiplying each element in the row of the left hand matrix by each corresponding element in the column of the right hand matrix and adding the results together.**
- **The number of columns in the left hand matrix must equal the number of rows in the right hand matrix.**
- **The product will then have the same number of rows as the left hand matrix and the same number of columns as the right hand matrix.**

So if

$$\mathbf{A} \times \mathbf{B} = \mathbf{C}$$

Dimensions: $(n \times m) \times (m \times k) \quad (n \times k)$

n is from the number of rows in **A**.
k is from the number of columns in **B**.

These numbers must be the same.

Example 6

Given that $\mathbf{A} = \begin{pmatrix} 1 & -2 \\ 3 & 4 \end{pmatrix}$ and $\mathbf{B} = \begin{pmatrix} -3 \\ 2 \end{pmatrix}$ find $\mathbf{AB}$.

First calculate the dimensions of AB

$(2 \times 2) \times (2 \times 1)$ gives (2×1)

The number of rows is two from here.

The number of columns is one from here.

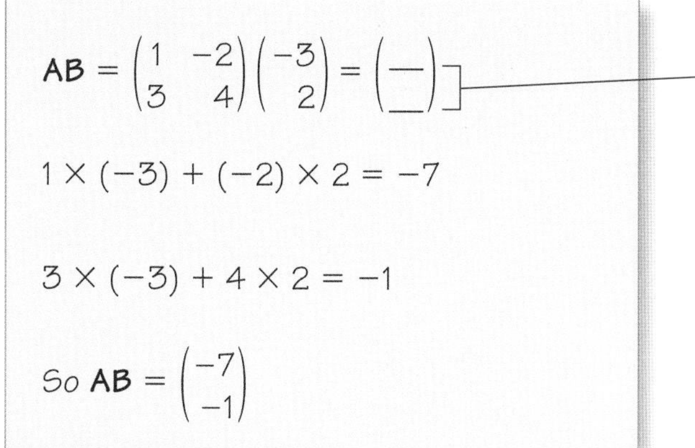

$AB = \begin{pmatrix} 1 & -2 \\ 3 & 4 \end{pmatrix}\begin{pmatrix} -3 \\ 2 \end{pmatrix} = \begin{pmatrix} \_ \\ \_ \end{pmatrix}$

$1 \times (-3) + (-2) \times 2 = -7$

$3 \times (-3) + 4 \times 2 = -1$

So $AB = \begin{pmatrix} -7 \\ -1 \end{pmatrix}$

Two numbers are required in the answer.

The top number is the total of the first row of **A** multiplied by the first column of **B**.

The bottom row is the total of the second row of **A** multiplied by the second column of **B**.

Example 7

Given that $\mathbf{A} = \begin{pmatrix} -1 & 0 \\ 2 & 3 \end{pmatrix}$ and $\mathbf{B} = \begin{pmatrix} 4 & 1 \\ 0 & -2 \end{pmatrix}$ find

a **AB**

b **BA**.

Matrices with the same number of rows and columns are called **square** matrices.
The product of two square matrices of the same size will always give a square matrix of the same size.

a **A** is a 2×2 matrix and **B** is a 2×2 matrix so they can be multiplied and the product will be a 2×2 matrix.

$AB = \begin{pmatrix} -1 & 0 \\ 2 & 3 \end{pmatrix}\begin{pmatrix} 4 & 1 \\ 0 & -2 \end{pmatrix} = \begin{pmatrix} a & b \\ c & d \end{pmatrix}$

$a = (-1) \times 4 + 0 \times 0 = -4$

$b = (-1) \times 1 + 0 \times (-2) = -1$

$c = 2 \times 4 + 3 \times 0 = 8$

$d = 2 \times 1 + 3 \times (-2) = -4$

So $AB = \begin{pmatrix} -4 & -1 \\ 8 & -4 \end{pmatrix}$

This time there are four numbers to be found.

a is the total of the first row multiplied by the first column.

b is the total of the first row multiplied by the second column.

c is the total of the second row multiplied by the first column.

d is the total of the second row multiplied by the second column.

b **BA** will also be a 2 × 2 matrix

$\begin{pmatrix} 4 & 1 \\ 0 & -2 \end{pmatrix}\begin{pmatrix} -1 & 0 \\ 2 & 3 \end{pmatrix} = \begin{pmatrix} -2 & 3 \\ -4 & -6 \end{pmatrix}$

First row times first column
$4 \times (-1) + 1 \times 2 = = -2$

First row times second column
$4 \times 0 + 1 \times 3 = 3$

Second row times second column
$0 \times 0 + (-2) \times 3 = -6$

Second row times first column
$0 \times (-1) + (-2) \times 2 = -4$

Notice that in Example 6 you could not find **BA** because the dimensions do not allow you to use the rule. In the case of square matrices though, like Example 7, you will always be able to find both **AB** and **BA**, and usually these answers will be different. (The technical term for this is to say that matrix multiplication is not **commutative**.) This means that, when you are finding the product of two matrices, it is very important that you place them in the correct order.

Example 8

Given that $\mathbf{A} = (1 \quad -1 \quad 2)$, $\mathbf{B} = (3 \quad -2)$ and $\mathbf{C} = \begin{pmatrix} 4 \\ 5 \end{pmatrix}$, determine whether or not the following products can be evaluated and, if they can, find the product.

a AB **b BC** **c AC** **d CA** **e BCA**

a $\mathbf{A} \times \mathbf{B}$

Dimensions: $(1 \times 3) \times (1 \times 2)$

Not possible

These two numbers are not the same so the product can not be found.

b $\mathbf{B} \times \mathbf{C}$

Dimensions: $(1 \times 2) \times (2 \times 1)$

These two numbers are the same so the product is possible and the answer will be 1×1.

$$\mathbf{BC} = (3 \quad -2)\begin{pmatrix} 4 \\ 5 \end{pmatrix} = (2)$$

Row times column: $3 \times 4 + (-2) \times 5 = 2$.

NB Matrix **B** is written $(3 \; -2)$ i.e. without a comma, to avoid confusion with the coordinates of a point $(3, -2)$.

c $\mathbf{A} \times \mathbf{C}$

Dimensions: $(1 \times 3) \times (2 \times 1)$

Not possible

These two numbers are not the same so the product can not be found.

d $\mathbf{C} \times \mathbf{A}$

Dimensions: $(2 \times 1) \times (1 \times 3)$

These two numbers are the same and the product will have dimension 2×3.

$$\mathbf{CA} = \begin{pmatrix} 4 \\ 5 \end{pmatrix}(1 \quad -1 \quad 2)$$

$$= \begin{pmatrix} 4 \times 1 & 4 \times (-1) & 4 \times 2 \\ 5 \times 1 & 5 \times (-1) & 4 \times 2 \end{pmatrix}$$

first row × first column gives $4 \times 1 = 4$
first row × second column gives $4 \times (-1) = -4$
first row × third column gives $4 \times 2 = 8$
second row × first column gives $5 \times 1 = 5$
second row × second column gives $5 \times (-1) = -5$
second row × third column gives $5 \times 2 = 10$

$$\text{So } \mathbf{CA} = \begin{pmatrix} 4 & -4 & 8 \\ 5 & -5 & 10 \end{pmatrix}$$

e $\mathbf{B} \times \mathbf{C} \times \mathbf{A}$

Dimensions: $(1 \times 2) \times (2 \times 1) \times (1 \times 3)$

These two numbers are the same AND these two numbers are the same. So the product is possible and will have dimension 1×3.

This can be calculated as $\mathbf{B}(\mathbf{CA})$ or $(\mathbf{BC})\mathbf{A}$. We shall do both.

$$\mathbf{B}(\mathbf{CA}) = (3 \quad -2)\begin{pmatrix} 4 & -4 & 8 \\ 5 & -5 & 10 \end{pmatrix} = (2 \quad -2 \quad 4)$$

Using **CA** from part **d**.

$$(\mathbf{BC})\mathbf{A} = (2)(1 \quad -1 \quad 2) = (2 \quad -2 \quad 4)$$

Using **BC** from part **b**.

- **When evaluating the product of three or more matrices, provided the order is kept the same it does not matter which product pair is evaluated first. (We say that matrix multiplication is associative.) Sometimes a little forward planning can make the evaluation simpler; (BC)A was easier than B(CA) in this case.**

- **Some calculators will evaluate matrix products and, whilst this facility may be useful to you in the FP1 examination, you should know how to calculate the product by hand as sometimes you will be given matrices containing unknown letters.**

Example 9

$\mathbf{A} = \begin{pmatrix} -1 \\ a \end{pmatrix}$ and $\mathbf{B} = (b \quad 2)$

Given that $\mathbf{BA} = (0)$, find $\mathbf{AB}$ in terms of a.

$$\mathbf{BA} = (b \quad 2)\begin{pmatrix} -1 \\ a \end{pmatrix} = (-b + 2a)$$

BA will be a 1×1matrix.

So $\mathbf{BA} = (0)$ implies that $b = 2a$

$$\mathbf{AB} = \begin{pmatrix} -1 \\ a \end{pmatrix}(b \quad 2) = \begin{pmatrix} -b & -2 \\ ab & 2a \end{pmatrix}$$

AB will be a 2×2 matrix.

Substituting $b = 2a$ gives $\mathbf{AB} = \begin{pmatrix} -2a & -2 \\ 2a^2 & 2a \end{pmatrix}$

Exercise 4C

1 Given the dimensions of the following matrices:

Matrix	**A**	**B**	**C**	**D**	**E**
Dimension	2×2	1×2	1×3	3×2	2×3

Give the dimensions of these matrix products.

a **BA** **b** **DE** **c** **CD**

d **ED** **e** **AE** **f** **DA**

2 Find these products.

a $\begin{pmatrix} 1 & 2 \\ 3 & 4 \end{pmatrix}\begin{pmatrix} -1 \\ 2 \end{pmatrix}$ **b** $\begin{pmatrix} 1 & 2 \\ 3 & 4 \end{pmatrix}\begin{pmatrix} 0 & 5 \\ -1 & -2 \end{pmatrix}$

3 The matrix $\mathbf{A} = \begin{pmatrix} -1 & -2 \\ 0 & 3 \end{pmatrix}$ and the matrix $\mathbf{B} = \begin{pmatrix} 1 & 0 & 1 \\ 1 & 1 & 0 \end{pmatrix}$.

Find

a **AB** **b** $\mathbf{A}^2$

$\mathbf{A}^2$ means $\mathbf{A} \times \mathbf{A}$

4 The matrices **A**, **B** and **C** are given by

$\mathbf{A} = \begin{pmatrix} 2 \\ 1 \end{pmatrix}$, $\mathbf{B} = \begin{pmatrix} 3 & 1 \\ -1 & 2 \end{pmatrix}$, $\mathbf{C} = (-3 \quad -2)$.

Determine whether or not the following products are possible and find the products of those that are.

a **AB** **b** **AC** **c** **BC**

d **BA** **e** **CA** **f** **CB**

5 Find in terms of a $\begin{pmatrix} 2 & a \\ 1 & -1 \end{pmatrix}\begin{pmatrix} 1 & 3 & 0 \\ 0 & -1 & 2 \end{pmatrix}$.

6 Find in terms of x $\begin{pmatrix} 3 & 2 \\ -1 & x \end{pmatrix}\begin{pmatrix} x & -2 \\ 1 & 3 \end{pmatrix}$.

7 The matrix $\mathbf{A} = \begin{pmatrix} 1 & 2 \\ 0 & 1 \end{pmatrix}$.

Find

a $\mathbf{A}^2$

b $\mathbf{A}^3$

c Suggest a form for $\mathbf{A}^k$.

You might be asked to prove this formula for $\mathbf{A}^k$ in FP1 using induction from Chapter 6.

8 The matrix $\mathbf{A} = \begin{pmatrix} a & 0 \\ b & 0 \end{pmatrix}$.

a Find, in terms of a and b, the matrix $\mathbf{A}^2$.

Given that $\mathbf{A}^2 = 3\mathbf{A}$

b find the value of a.

9 $\mathbf{A} = (-1 \quad 3)$, $\mathbf{B} = \begin{pmatrix} 2 \\ 1 \\ 0 \end{pmatrix}$, $\mathbf{C} = \begin{pmatrix} 4 & -2 \\ 0 & -3 \end{pmatrix}$.

Find **a** **BAC** **b** $\mathbf{AC}^2$

10 $\mathbf{A} = \begin{pmatrix} 1 \\ -1 \\ 2 \end{pmatrix}$, $\mathbf{B} = (3 \quad -2 \quad -3)$.

Find **a** **ABA** **b** **BAB**

4.5 You can use matrices to describe linear transformations.

- You may be familiar with the idea of a **transformation** from GCSE. There you may have met transformations such as **rotations, translations, reflections** and **enlargements.** There is more about these specific types of transformations in Section 4.6.
- A transformation moves all the points (x, y) in a plane according to some rule.
- You can describe a transformation in terms of its effect on the **position vector** $\begin{pmatrix} x \\ y \end{pmatrix}$, this is simply the **vector** from the point (0, 0) to the point (x, y).
- You call the new point to which $\begin{pmatrix} x \\ y \end{pmatrix}$ is moved the **image** of $\begin{pmatrix} x \\ y \end{pmatrix}$.

Example 10

The three transformations **S**, **T** and **U** are defined as follows. Find the images of the point (2, 3) under each of these transformations.

$$\mathbf{S}: \begin{pmatrix} x \\ y \end{pmatrix} \rightarrow \begin{pmatrix} x+4 \\ y-1 \end{pmatrix}, \quad \mathbf{T}: \begin{pmatrix} x \\ y \end{pmatrix} \rightarrow \begin{pmatrix} 2x-y \\ x+y \end{pmatrix}, \quad \mathbf{U}: \begin{pmatrix} x \\ y \end{pmatrix} \rightarrow \begin{pmatrix} 2y \\ -x^2 \end{pmatrix},$$

$\mathrm{T}\begin{pmatrix} 2 \\ 3 \end{pmatrix} = \begin{pmatrix} 2 \times 2 - 3 \\ 2 + 3 \end{pmatrix} = \begin{pmatrix} 1 \\ 5 \end{pmatrix}$ i.e. the point (1, 5)

Substituting $x = 2$ and $y = 3$ into $2x - y$ gives 1 and into $x + y$ gives 5.

$\mathrm{S}\begin{pmatrix} 2 \\ 3 \end{pmatrix} = \begin{pmatrix} 2 + 4 \\ 3 - 1 \end{pmatrix} = \begin{pmatrix} 6 \\ 2 \end{pmatrix}$ i.e. the point (6, 2)

Substitute $x = 2$ and $y = 3$ into the **S** formulae.

$\mathrm{U}\begin{pmatrix} 2 \\ 3 \end{pmatrix} = \begin{pmatrix} 2 \times 3 \\ -2^2 \end{pmatrix} = \begin{pmatrix} 6 \\ -4 \end{pmatrix}$ i.e. the point (6, −4)

N.B. when $x = 2$, $-x^2 = -2^2 = -4$

- A **linear transformation** has the special properties that the transformation only involves linear expressions of x and y (so **U** in Example 10 is not linear) and that the origin (0, 0) is not moved by the transformation (so **S** in Example 10 is not linear).

- **A linear transformation T has these properties.**

  1 $\mathbf{T}\begin{pmatrix} kx \\ ky \end{pmatrix} = k\mathbf{T}\begin{pmatrix} x \\ y \end{pmatrix}$, **where k is a constant.**

  2 $\mathbf{T}\left[\begin{pmatrix} x_1 \\ y_1 \end{pmatrix} + \begin{pmatrix} x_2 \\ y_2 \end{pmatrix}\right] = \mathbf{T}\begin{pmatrix} x_1 \\ y_1 \end{pmatrix} + \mathbf{T}\begin{pmatrix} x_2 \\ y_2 \end{pmatrix}$

Example 11 illustrates how the definitions of linear transformations work and is not typical of questions you will find in the FP1 examination.

Example 11

For the transformations **S**, **T** and **U** from Example 10, use the vectors $\begin{pmatrix} 1 \\ -1 \end{pmatrix}$ and $\begin{pmatrix} -2 \\ 1 \end{pmatrix}$ to show that **S** and **U** do not satisfy property 2 and prove that **T** satisfies both properties.

Using **S**

$$\text{LHS} = \mathbf{S}\left[\begin{pmatrix} 1 \\ -1 \end{pmatrix} + \begin{pmatrix} -2 \\ 1 \end{pmatrix}\right] = \mathbf{S}\begin{pmatrix} -1 \\ 0 \end{pmatrix} = \begin{pmatrix} -1+4 \\ 0-1 \end{pmatrix} = \begin{pmatrix} 3 \\ -1 \end{pmatrix}$$

$$\text{RHS} = \mathbf{S}\begin{pmatrix} 1 \\ -1 \end{pmatrix} + \mathbf{S}\begin{pmatrix} -2 \\ 1 \end{pmatrix} = \begin{pmatrix} 1+4 \\ -1-1 \end{pmatrix} + \begin{pmatrix} -2+4 \\ 1-1 \end{pmatrix} = \begin{pmatrix} 5 \\ -2 \end{pmatrix} + \begin{pmatrix} 2 \\ 0 \end{pmatrix} = \begin{pmatrix} 7 \\ -2 \end{pmatrix}$$

These two points are not the same so **S** is not linear.

So **S** is not a linear transformation.

Using **U**

$$\text{LHS} = \mathbf{U}\left[\begin{pmatrix} 1 \\ -1 \end{pmatrix} + \begin{pmatrix} -2 \\ 1 \end{pmatrix}\right] = \mathbf{U}\begin{pmatrix} -1 \\ 0 \end{pmatrix} = \begin{pmatrix} 2 \times 0 \\ -(-1)^2 \end{pmatrix} = \begin{pmatrix} 0 \\ -1 \end{pmatrix}$$

$$\text{RHS} = \mathbf{U}\begin{pmatrix} 1 \\ -1 \end{pmatrix} + \mathbf{U}\begin{pmatrix} -2 \\ 1 \end{pmatrix} = \begin{pmatrix} 2 \times -1 \\ -1^2 \end{pmatrix} + \begin{pmatrix} 2 \times 1 \\ -(-2)^2 \end{pmatrix} = \begin{pmatrix} -2 \\ -1 \end{pmatrix} + \begin{pmatrix} 2 \\ -4 \end{pmatrix} = \begin{pmatrix} 0 \\ -5 \end{pmatrix}$$

These are not the same so **U** is not a linear transformation.

To prove that **T** is a linear transformation you need to use general points, not specific vectors like we did for **S** and **U**.

$$\mathbf{T}\left[\begin{pmatrix} x \\ y \end{pmatrix}\right] = \mathbf{T}\begin{pmatrix} kx \\ ky \end{pmatrix} = \begin{pmatrix} 2kx - ky \\ kx + ky \end{pmatrix}$$

$$= k\begin{pmatrix} 2x - y \\ x + y \end{pmatrix} = k\mathbf{T}\begin{pmatrix} x \\ y \end{pmatrix}$$

The rule $\begin{pmatrix} x \\ y \end{pmatrix} \to \begin{pmatrix} 2x - y \\ x + y \end{pmatrix}$ means that the new top number is twice the old top number minus the old bottom number.

So **T** satisfies property 1.

$$\mathbf{T}\left[\begin{pmatrix} x_1 \\ y_1 \end{pmatrix} + \begin{pmatrix} x_2 \\ y_2 \end{pmatrix}\right] = \mathbf{T}\begin{pmatrix} x_1 + x_2 \\ y_1 + y_2 \end{pmatrix} = \begin{pmatrix} 2[x_1 + x_2] - [y_1 + y_2] \\ x_1 + x_2 + [y_1 + y_2] \end{pmatrix}$$

$$\mathbf{T}\begin{pmatrix} x_1 \\ y_1 \end{pmatrix} + \mathbf{T}\begin{pmatrix} x_2 \\ y_2 \end{pmatrix} = \begin{pmatrix} 2x_1 - y_1 \\ x_1 + y_1 \end{pmatrix} + \begin{pmatrix} 2x_2 - y_2 \\ x_2 + y_2 \end{pmatrix} = \begin{pmatrix} 2[x_1 + x_2] - [y_1 + y_2] \\ x_1 + x_2 + [y_1 + y_2] \end{pmatrix}$$

Compare LHS and RHS of property 2 using the rule for **T**.

So **T** satisfies property 2.

Therefore **T** is a linear transformation.

- **Any linear transformation can be represented by a matrix.**
- **The linear transformation $\mathbf{T}: \begin{pmatrix} x \\ y \end{pmatrix} \rightarrow \begin{pmatrix} ax + by \\ cx + dy \end{pmatrix}$ can be represented by the matrix $\mathbf{M} = \begin{pmatrix} a & b \\ c & d \end{pmatrix}$**

since $\begin{pmatrix} a & b \\ c & d \end{pmatrix}\begin{pmatrix} x \\ y \end{pmatrix} = \begin{pmatrix} ax + by \\ cx + dy \end{pmatrix}$.

Example 12

Find matrices to represent these linear transformations.

a $\mathbf{T}: \begin{pmatrix} x \\ y \end{pmatrix} \rightarrow \begin{pmatrix} 2y + x \\ 3x \end{pmatrix}$ **b** $\mathbf{V}: \begin{pmatrix} x \\ y \end{pmatrix} \rightarrow \begin{pmatrix} -2y \\ 3x + y \end{pmatrix}$

a Transformation **T** is equivalent to

$\mathbf{T}: \begin{pmatrix} x \\ y \end{pmatrix} \rightarrow \begin{pmatrix} 1x + 2y \\ 3x + 0y \end{pmatrix}$

Write the transformation in the form $\begin{pmatrix} ax + by \\ cx + dy \end{pmatrix}$.

so the matrix is $\begin{pmatrix} 1 & 2 \\ 3 & 0 \end{pmatrix}$.

Use the coefficients of x and y to form the matrix.

b Transformation **V** is equivalent to

$\mathbf{V}: \begin{pmatrix} x \\ y \end{pmatrix} \rightarrow \begin{pmatrix} 0x - 2y \\ 3x + y \end{pmatrix}$

Write the transformation in the form $\begin{pmatrix} ax + by \\ cx + dy \end{pmatrix}$.

so the matrix is $\begin{pmatrix} 0 & -2 \\ 3 & 1 \end{pmatrix}$.

Use the coefficients of x and y to form the matrix.

Example 13

The square S has coordinates (1, 1), (3, 1), (3, 3) and (1, 3). Find the vertices of the image of S under the transformation given by the matrix $\mathbf{M} = \begin{pmatrix} -1 & 2 \\ 2 & 1 \end{pmatrix}$.

The vertices of S can be represented by the matrix

$\begin{pmatrix} 1 & 3 & 3 & 1 \\ 1 & 1 & 3 & 3 \end{pmatrix}$

Write each point as a column vector so (1, 3) becomes $\begin{pmatrix} 1 \\ 3 \end{pmatrix}$.

Then combine all four column vectors into a single 2×4 matrix.

To find the image of S you need the matrix product:

$\begin{pmatrix} -1 & 2 \\ 2 & 1 \end{pmatrix}\begin{pmatrix} 1 & 3 & 3 & 1 \\ 1 & 1 & 3 & 3 \end{pmatrix}$

Use the usual rule for multiplying matrices.

$= \begin{pmatrix} 1 & -1 & 3 & 5 \\ 3 & 7 & 9 & 5 \end{pmatrix}$

e.g. first row and fourth column: $-1 \times 1 + 2 \times 3 = 5$

e.g. second row and third column: $2 \times 3 + 1 \times 3 = 9$

Exercise 4D

1 Which of the following are not linear transformations?

a $\mathbf{P}: \begin{pmatrix} x \\ y \end{pmatrix} \rightarrow \begin{pmatrix} 2x \\ y + 1 \end{pmatrix}$

b $\mathbf{Q}: \begin{pmatrix} x \\ y \end{pmatrix} \rightarrow \begin{pmatrix} x^2 \\ y \end{pmatrix}$

c $\mathbf{R}: \begin{pmatrix} x \\ y \end{pmatrix} \rightarrow \begin{pmatrix} 2x + y \\ x + xy \end{pmatrix}$

d $\mathbf{S}: \begin{pmatrix} x \\ y \end{pmatrix} \rightarrow \begin{pmatrix} 3y \\ -x \end{pmatrix}$

e $\mathbf{T}: \begin{pmatrix} x \\ y \end{pmatrix} \rightarrow \begin{pmatrix} y + 3 \\ x + 3 \end{pmatrix}$

f $\mathbf{U}: \begin{pmatrix} x \\ y \end{pmatrix} \rightarrow \begin{pmatrix} 2x \\ 3y - 2x \end{pmatrix}$

2 Identify which of these are linear transformations and give their matrix representations. Give reasons to explain why the other transformations are not linear.

a $\mathbf{S}: \begin{pmatrix} x \\ y \end{pmatrix} \rightarrow \begin{pmatrix} 2x - y \\ 3x \end{pmatrix}$

b $\mathbf{T}: \begin{pmatrix} x \\ y \end{pmatrix} \rightarrow \begin{pmatrix} 2y + 1 \\ x - 1 \end{pmatrix}$

c $\mathbf{U}: \begin{pmatrix} x \\ y \end{pmatrix} \rightarrow \begin{pmatrix} xy \\ 0 \end{pmatrix}$

d $\mathbf{V}: \begin{pmatrix} x \\ y \end{pmatrix} \rightarrow \begin{pmatrix} 2y \\ -x \end{pmatrix}$

e $\mathbf{W}: \begin{pmatrix} x \\ y \end{pmatrix} \rightarrow \begin{pmatrix} y \\ x \end{pmatrix}$

3 Identify which of these are linear transformations and give their matrix representations. Give reasons to explain why the other transformations are not linear.

a $\mathbf{S}: \begin{pmatrix} x \\ y \end{pmatrix} \rightarrow \begin{pmatrix} x^2 \\ y^2 \end{pmatrix}$

b $\mathbf{T}: \begin{pmatrix} x \\ y \end{pmatrix} \rightarrow \begin{pmatrix} -y \\ x \end{pmatrix}$

c $\mathbf{U}: \begin{pmatrix} x \\ y \end{pmatrix} \rightarrow \begin{pmatrix} x - y \\ x - y \end{pmatrix}$

d $\mathbf{V}: \begin{pmatrix} x \\ y \end{pmatrix} \rightarrow \begin{pmatrix} 0 \\ 0 \end{pmatrix}$

e $\mathbf{W}: \begin{pmatrix} x \\ y \end{pmatrix} \rightarrow \begin{pmatrix} x \\ y \end{pmatrix}$

4 Find matrix representations for these linear transformations.

a $\begin{pmatrix} x \\ y \end{pmatrix} \rightarrow \begin{pmatrix} y + 2x \\ -y \end{pmatrix}$

b $\begin{pmatrix} x \\ y \end{pmatrix} \rightarrow \begin{pmatrix} -y \\ x + 2y \end{pmatrix}$

5 The triangle T has vertices at $(-1, 1)$, $(2, 3)$ and $(5, 1)$.
Find the vertices of the image of T under the transformations represented by these matrices.

a $\begin{pmatrix} -1 & 0 \\ 0 & 1 \end{pmatrix}$

b $\begin{pmatrix} 1 & 4 \\ 0 & -2 \end{pmatrix}$

c $\begin{pmatrix} 0 & -2 \\ 2 & 0 \end{pmatrix}$

6 The square S has vertices at $(-1, 0)$, $(0, 1)$, $(1, 0)$ and $(0, -1)$.

Find the vertices of the image of S under the transformations represented by these matrices.

a $\begin{pmatrix} 2 & 0 \\ 0 & 3 \end{pmatrix}$ **b** $\begin{pmatrix} 1 & -1 \\ 1 & 1 \end{pmatrix}$ **c** $\begin{pmatrix} 1 & 1 \\ 1 & -1 \end{pmatrix}$

4.6 You can use matrices to represent rotations, reflections and enlargements.

- **In GCSE you may have met some simple transformations such as rotations, reflections, enlargements and translations.**

- **A translation is *not* a linear transformation (since the origin moves) but all the others are and in this section we shall see how to represent them using matrices.**

- **To identify the matrix representing a particular transformation you should consider the effect of the matrix or the transformation on two simple vectors $\begin{pmatrix} 1 \\ 0 \end{pmatrix}$ (sometimes denoted as i) and $\begin{pmatrix} 0 \\ 1 \end{pmatrix}$ (sometimes denoted by j).**

y, $\begin{pmatrix} 0 \\ 1 \end{pmatrix}$, O, $\begin{pmatrix} 1 \\ 0 \end{pmatrix}$, x

- **Given any matrix $\mathbf{M} = \begin{pmatrix} a & b \\ c & d \end{pmatrix}$ you can see that**

$$\mathbf{M}\begin{pmatrix} 1 \\ 0 \end{pmatrix} = \begin{pmatrix} a \\ c \end{pmatrix} \quad \text{and} \quad \mathbf{M}\begin{pmatrix} 0 \\ 1 \end{pmatrix} = \begin{pmatrix} b \\ d \end{pmatrix}$$

so the first column of M gives the image of $\begin{pmatrix} 1 \\ 0 \end{pmatrix}$ and the second column of M gives the image of $\begin{pmatrix} 0 \\ 1 \end{pmatrix}$.

- **You can use this information to identify the transformation represented by a matrix.**

In FP1 you should be able to identify matrices representing the following linear transformations.

Rotation	about (0, 0) of angles that are multiples of 45°.
Enlargement	centre (0, 0) of scale factor $k (k \neq 0, k \in \mathbb{R})$.
Reflection	in coordinate axes or the lines $y = \pm x$.
Identity	the matrix $\begin{pmatrix} 1 & 0 \\ 0 & 1 \end{pmatrix}$ is called **I** and does not carry out any transformation. (This is equivalent to multiplying by 1 in arithmetic.)

Example 14

Describe fully the geometrical transformations represented by these matrices.

a $\begin{pmatrix} 3 & 0 \\ 0 & 3 \end{pmatrix}$ **b** $\begin{pmatrix} -1 & 0 \\ 0 & -1 \end{pmatrix}$ **c** $\begin{pmatrix} 0 & -1 \\ -1 & 0 \end{pmatrix}$

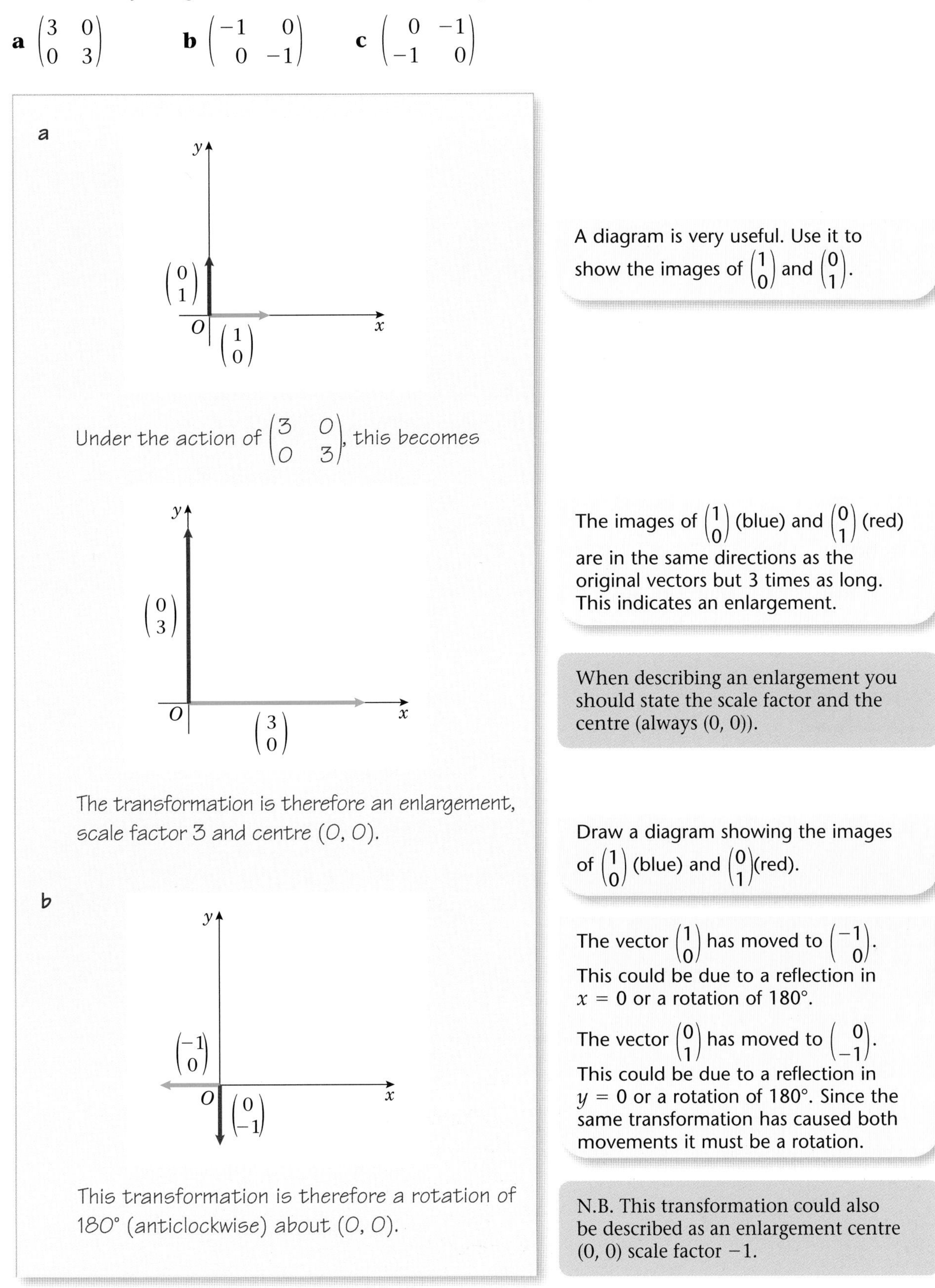

a

A diagram is very useful. Use it to show the images of $\begin{pmatrix} 1 \\ 0 \end{pmatrix}$ and $\begin{pmatrix} 0 \\ 1 \end{pmatrix}$.

Under the action of $\begin{pmatrix} 3 & 0 \\ 0 & 3 \end{pmatrix}$, this becomes

The images of $\begin{pmatrix} 1 \\ 0 \end{pmatrix}$ (blue) and $\begin{pmatrix} 0 \\ 1 \end{pmatrix}$ (red) are in the same directions as the original vectors but 3 times as long. This indicates an enlargement.

When describing an enlargement you should state the scale factor and the centre (always (0, 0)).

The transformation is therefore an enlargement, scale factor 3 and centre (0, 0).

b

Draw a diagram showing the images of $\begin{pmatrix} 1 \\ 0 \end{pmatrix}$ (blue) and $\begin{pmatrix} 0 \\ 1 \end{pmatrix}$(red).

The vector $\begin{pmatrix} 1 \\ 0 \end{pmatrix}$ has moved to $\begin{pmatrix} -1 \\ 0 \end{pmatrix}$. This could be due to a reflection in $x = 0$ or a rotation of 180°.

The vector $\begin{pmatrix} 0 \\ 1 \end{pmatrix}$ has moved to $\begin{pmatrix} 0 \\ -1 \end{pmatrix}$. This could be due to a reflection in $y = 0$ or a rotation of 180°. Since the same transformation has caused both movements it must be a rotation.

This transformation is therefore a rotation of 180° (anticlockwise) about (0, 0).

N.B. This transformation could also be described as an enlargement centre (0, 0) scale factor −1.

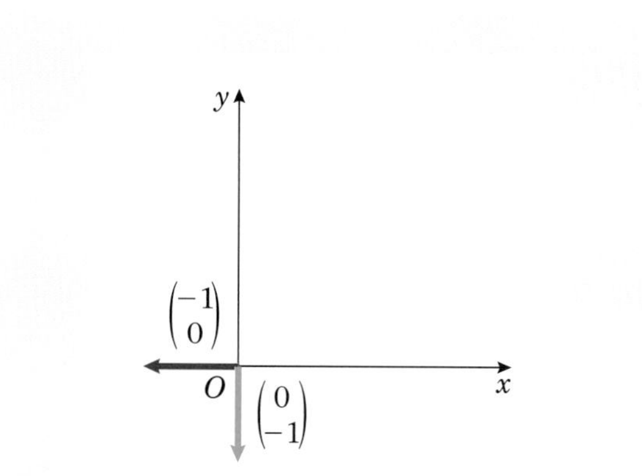

This transformation is therefore a reflection in the line $y = -x$.

When describing a rotation you should state the angle, direction (anticlockwise is positive) and centre (always (0, 0)).

The diagram shows that $\begin{pmatrix}1\\0\end{pmatrix}$ has moved to $\begin{pmatrix}0\\-1\end{pmatrix}$. This could be due to a rotation of $-90°$or a reflection in $y = -x$.

Also $\begin{pmatrix}0\\1\end{pmatrix}$ has moved to $\begin{pmatrix}-1\\0\end{pmatrix}$. This could be due to a rotation of $+90°$or a reflection in $y = -x$. Since the same transformation has carried out both movements it must be a reflection.

To describe a reflection you must state the mirror line.

Parts **b** and **c** of this example are often confused so a carefully drawn diagram is useful.

Another way of identifying a reflection is to look at the determinant of the matrix (see Section 4.8).

If this is negative the transformation involves a reflection.

If $\mathbf{M} = \begin{pmatrix}a & b\\c & d\end{pmatrix}$ the determinant of $\mathbf{M}$ is given by $ad - bc$.

Example 15

Find matrices to represent these transformations.

a Rotation of 45° anticlockwise about (0, 0).

b Reflection in the y-axis.

c Enlargement centre (0, 0) of scale factor -2.

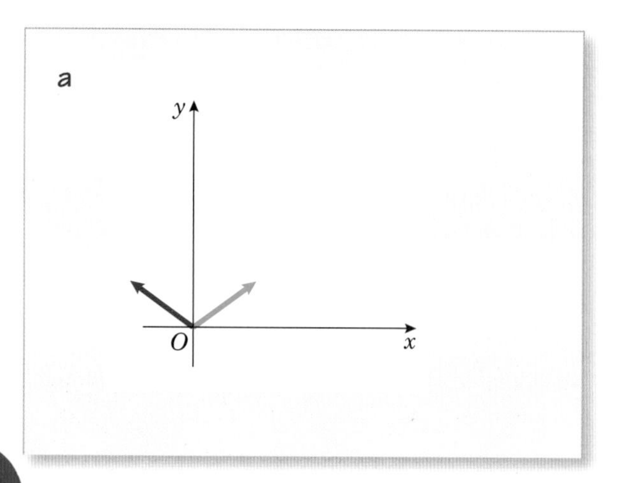

Draw a diagram showing the images of the original vectors after the rotation of 45°.

Consider the image of $\begin{pmatrix} 1 \\ 0 \end{pmatrix}$.

1

sin 45° $= \frac{1}{\sqrt{2}}$

45°

cos 45° $= \frac{1}{\sqrt{2}}$

Remember that cos 45° $= \dfrac{1}{\sqrt{2}}$ and sin 45° $= \dfrac{1}{\sqrt{2}}$.

This will be $\begin{pmatrix} \frac{1}{\sqrt{2}} \\ \frac{1}{\sqrt{2}} \end{pmatrix}$.

Similarly the image of $\begin{pmatrix} 0 \\ 1 \end{pmatrix}$ is $\begin{pmatrix} -\frac{1}{\sqrt{2}} \\ \frac{1}{\sqrt{2}} \end{pmatrix}$.

Therefore the matrix representing this transformation

is $\begin{pmatrix} \frac{1}{\sqrt{2}} & -\frac{1}{\sqrt{2}} \\ \frac{1}{\sqrt{2}} & \frac{1}{\sqrt{2}} \end{pmatrix}$

Remember that the image of $\begin{pmatrix} 1 \\ 0 \end{pmatrix}$ gives the first column and the image of $\begin{pmatrix} 0 \\ 1 \end{pmatrix}$ the second.

b

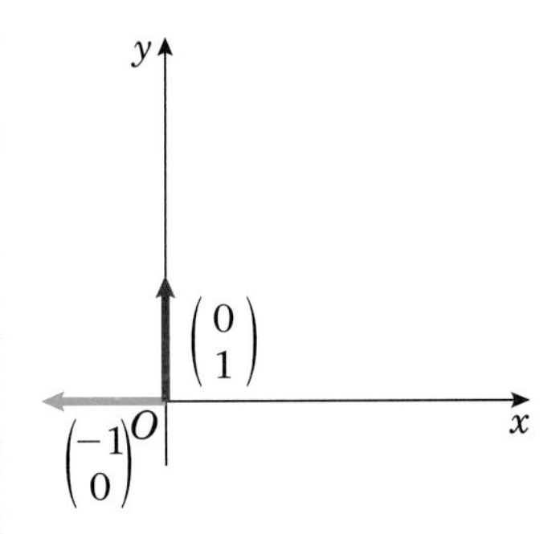

The diagram shows that $\begin{pmatrix} 1 \\ 0 \end{pmatrix}$ moves to $\begin{pmatrix} -1 \\ 0 \end{pmatrix}$ but $\begin{pmatrix} 0 \\ 1 \end{pmatrix}$ remains unchanged.

The matrix representing a reflection in the line $x = 0$ is $\begin{pmatrix} -1 & 0 \\ 0 & 1 \end{pmatrix}$.

The first column is $\begin{pmatrix} -1 \\ 0 \end{pmatrix}$, the image of $\begin{pmatrix} 1 \\ 0 \end{pmatrix}$ and the second column is $\begin{pmatrix} 0 \\ 1 \end{pmatrix}$, the image of $\begin{pmatrix} 0 \\ 1 \end{pmatrix}$.

c Image of $\begin{pmatrix} 1 \\ 0 \end{pmatrix}$ is $-2 \times \begin{pmatrix} 1 \\ 0 \end{pmatrix} = \begin{pmatrix} -2 \\ 0 \end{pmatrix}$

Image of $\begin{pmatrix} 0 \\ 1 \end{pmatrix}$ is $-2 \times \begin{pmatrix} 0 \\ 1 \end{pmatrix} = \begin{pmatrix} 0 \\ -2 \end{pmatrix}$

To enlarge a vector by scale factor k you simply multiply the vector by k.

The matrix representing an enlargement of scale factor -2 about $(0, 0)$ is $\begin{pmatrix} -2 & 0 \\ 0 & -2 \end{pmatrix}$.

Image of $\begin{pmatrix} 1 \\ 0 \end{pmatrix}$ as the first column and the image of $\begin{pmatrix} 0 \\ 1 \end{pmatrix}$ as the second.

Exercise 4E

1 Describe fully the geometrical transformations represented by these matrices.

a $\begin{pmatrix} 1 & 0 \\ 0 & -1 \end{pmatrix}$ **b** $\begin{pmatrix} 0 & -1 \\ 1 & 0 \end{pmatrix}$ **c** $\begin{pmatrix} 0 & 1 \\ -1 & 0 \end{pmatrix}$

2 Describe fully the geometrical transformations represented by these matrices.

a $\begin{pmatrix} \frac{1}{2} & 0 \\ 0 & \frac{1}{2} \end{pmatrix}$ **b** $\begin{pmatrix} 0 & 1 \\ 1 & 0 \end{pmatrix}$ **c** $\begin{pmatrix} 1 & 0 \\ 0 & 1 \end{pmatrix}$

3 Describe fully the geometrical transformations represented by these matrices.

a $\begin{pmatrix} \frac{1}{\sqrt{2}} & \frac{1}{\sqrt{2}} \\ -\frac{1}{\sqrt{2}} & \frac{1}{\sqrt{2}} \end{pmatrix}$ **b** $\begin{pmatrix} 4 & 0 \\ 0 & 4 \end{pmatrix}$ **c** $\frac{1}{\sqrt{2}}\begin{pmatrix} -1 & 1 \\ -1 & -1 \end{pmatrix}$

4 Find the matrix that represents these transformations.

a Rotation of 90° clockwise about (0, 0).

b Reflection in the x-axis.

c Enlargement centre (0, 0) scale factor 2.

5 Find the matrix that represents these transformations.

a Enlargement scale factor -4 centre (0, 0).

b Reflection in the line $y = x$.

c Rotation about (0, 0) of 135° anticlockwise.

4.7 You can use matrix products to represent combinations of transformations.

Example 16

The points $A(1, 0)$, $B(0, 1)$ and $C(2, 0)$ are the vertices of a triangle T. The triangle T is rotated through 90° anticlockwise about (0, 0) and then the image T' is reflected in the line $y = x$ to obtain the triangle T''.

a On separate diagrams sketch T, T' and T''.

b Find the matrix **P** such that $\mathbf{P}(T) = T'$ and the matrix **Q** such that $\mathbf{Q}(T') = T''$.

c By forming a suitable matrix product find the matrix **R** such that $\mathbf{R}(T) = T''$.

d Describe the single geometrical transformation represented by **R**.

e Find the single matrix representing a reflection in the line $y = x$ followed by a rotation of 90° anticlockwise about (0, 0).

f Describe the geometrical effect of the matrix found in part **e** as a single transformation.

a

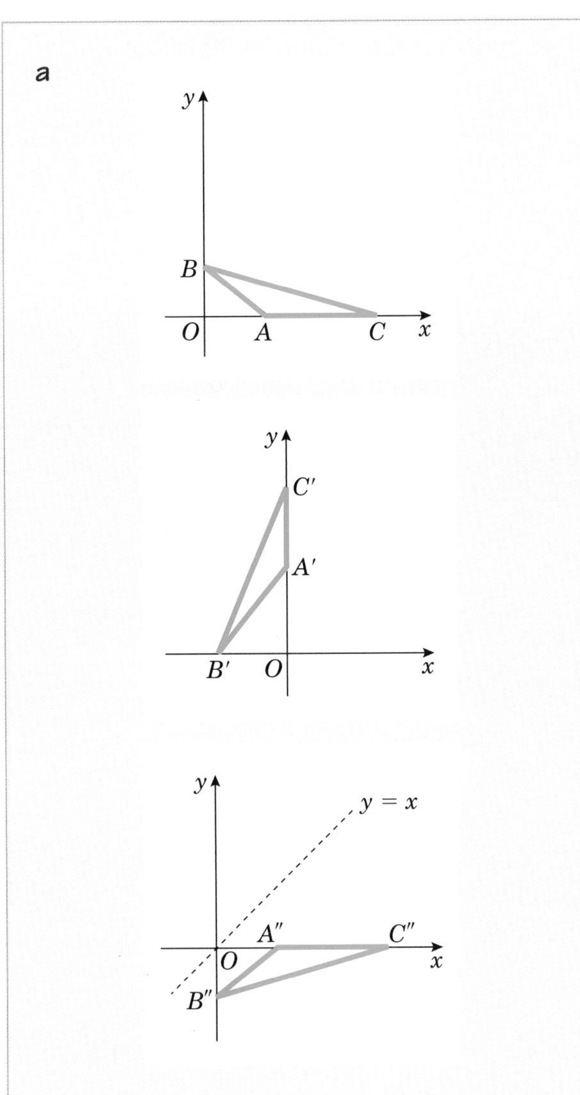

This is the original triangle T.

Rotate 90° about (0, 0) so that $\begin{pmatrix}1\\0\end{pmatrix}$ moves to $\begin{pmatrix}0\\1\end{pmatrix}$ and $\begin{pmatrix}0\\1\end{pmatrix}$ moves to $\begin{pmatrix}-1\\0\end{pmatrix}$.
This gives T'.

Reflect T' in $y = x$ so that $\begin{pmatrix}0\\1\end{pmatrix}$ moves to $\begin{pmatrix}1\\0\end{pmatrix}$ and $\begin{pmatrix}-1\\0\end{pmatrix}$ moves to $\begin{pmatrix}0\\-1\end{pmatrix}$.
This gives T''.

b Note that OA moves to OA' and then to OA'' and OB moves to OB' and then to OB''.

P represents a rotation of 90° about (0, 0) and so

$$\mathbf{P} = \begin{pmatrix}0 & -1\\1 & 0\end{pmatrix}$$

$\begin{pmatrix}1\\0\end{pmatrix} \rightarrow \begin{pmatrix}0\\1\end{pmatrix}$ and $\begin{pmatrix}0\\1\end{pmatrix} \rightarrow \begin{pmatrix}-1\\0\end{pmatrix}$

Q represents a reflection in $y = x$ so

$$\mathbf{Q} = \begin{pmatrix}0 & 1\\1 & 0\end{pmatrix}$$

$\begin{pmatrix}1\\0\end{pmatrix} \rightarrow \begin{pmatrix}0\\1\end{pmatrix}$ and $\begin{pmatrix}0\\1\end{pmatrix} \rightarrow \begin{pmatrix}1\\0\end{pmatrix}$

c $\mathbf{P}(T) = T'$ and $\mathbf{Q}(T') = T''$
So $\mathbf{Q}(\mathbf{P}(T)) = T''$
And therefore $\mathbf{QP} = \mathbf{R}$

Substitute $\mathbf{P}(T)$ for T'.
$\mathbf{Q}(\mathbf{P}(T))$ can be written as $\mathbf{QP}(T)$

$$\mathbf{R} = \mathbf{QP} = \begin{pmatrix} 0 & 1 \\ 1 & 0 \end{pmatrix}\begin{pmatrix} 0 & -1 \\ 1 & 0 \end{pmatrix}$$

$$= \begin{pmatrix} 1 & 0 \\ 0 & -1 \end{pmatrix}$$

Use the usual rules of matrix multiplication met in Section 4.4.

d

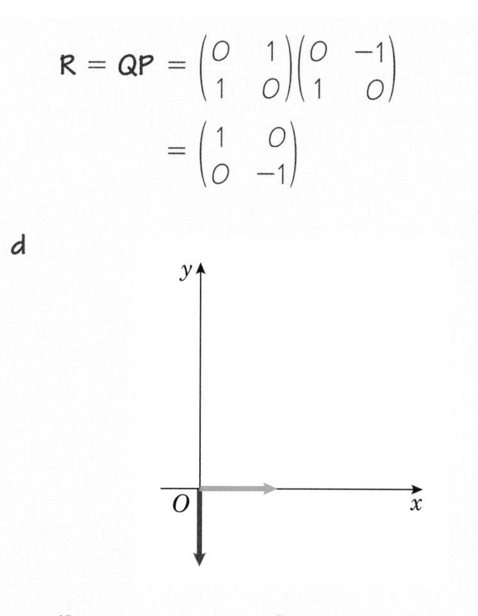

Draw a diagram showing the images of $\begin{pmatrix} 1 \\ 0 \end{pmatrix}$ (blue) and $\begin{pmatrix} 0 \\ 1 \end{pmatrix}$ (red) under **R**.

R represents a refection in the x-axis or the line y = 0

$\begin{pmatrix} 1 \\ 0 \end{pmatrix}$ is unmoved (since it lies on the mirror line) and $\begin{pmatrix} 0 \\ 1 \end{pmatrix}$ moves to $\begin{pmatrix} 0 \\ -1 \end{pmatrix}$ which confirms the reflection.

e The transformations are the other way around. The required matrix is **PQ**.

$$\mathbf{PQ} = \begin{pmatrix} 0 & -1 \\ 1 & 0 \end{pmatrix}\begin{pmatrix} 0 & 1 \\ 1 & 0 \end{pmatrix}$$

$$= \begin{pmatrix} -1 & 0 \\ 0 & 1 \end{pmatrix}$$

To transform with **Q** first followed by **P** you calculate **PQ**.

f

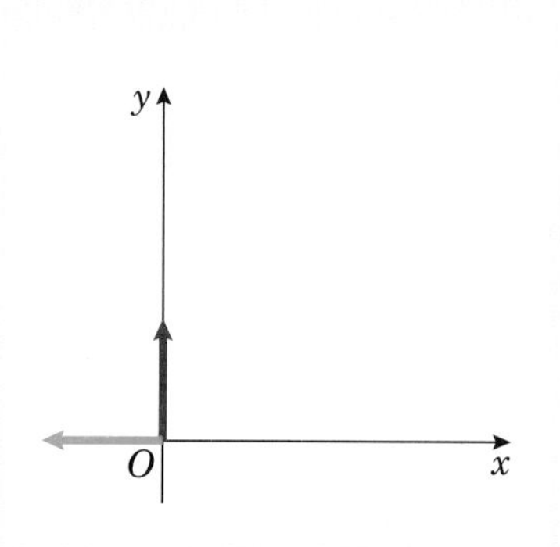

Draw a diagram showing the images of $\begin{pmatrix} 1 \\ 0 \end{pmatrix}$ (blue) and $\begin{pmatrix} 0 \\ 1 \end{pmatrix}$ (red).

PQ represents a reflection in the y-axis (or the line $x = 0$).

In general: the matrix **PQ** represents the transformation represented by **Q** followed by the transformation represented by **P**.

You can combine more than two transformations in a similar fashion.

The fact that $\begin{pmatrix} 0 \\ 1 \end{pmatrix}$ does not move suggests that the y-axis is a mirror line. Consideration of $\begin{pmatrix} 1 \\ 0 \end{pmatrix}$ confirms this.

Example 17

The matrices $\mathbf{P} = \begin{pmatrix} 1 & 1 \\ 2 & 3 \end{pmatrix}$, $\mathbf{Q} = \begin{pmatrix} 1 & 2 \\ 0 & 1 \end{pmatrix}$ and $\mathbf{R} = \begin{pmatrix} 3 & 7 \\ -1 & -2 \end{pmatrix}$ represent three transformations.

Find the matrix representing the transformation represented by **R**, followed by **Q** followed by **P** and give a geometrical interpretation of this combined transformation.

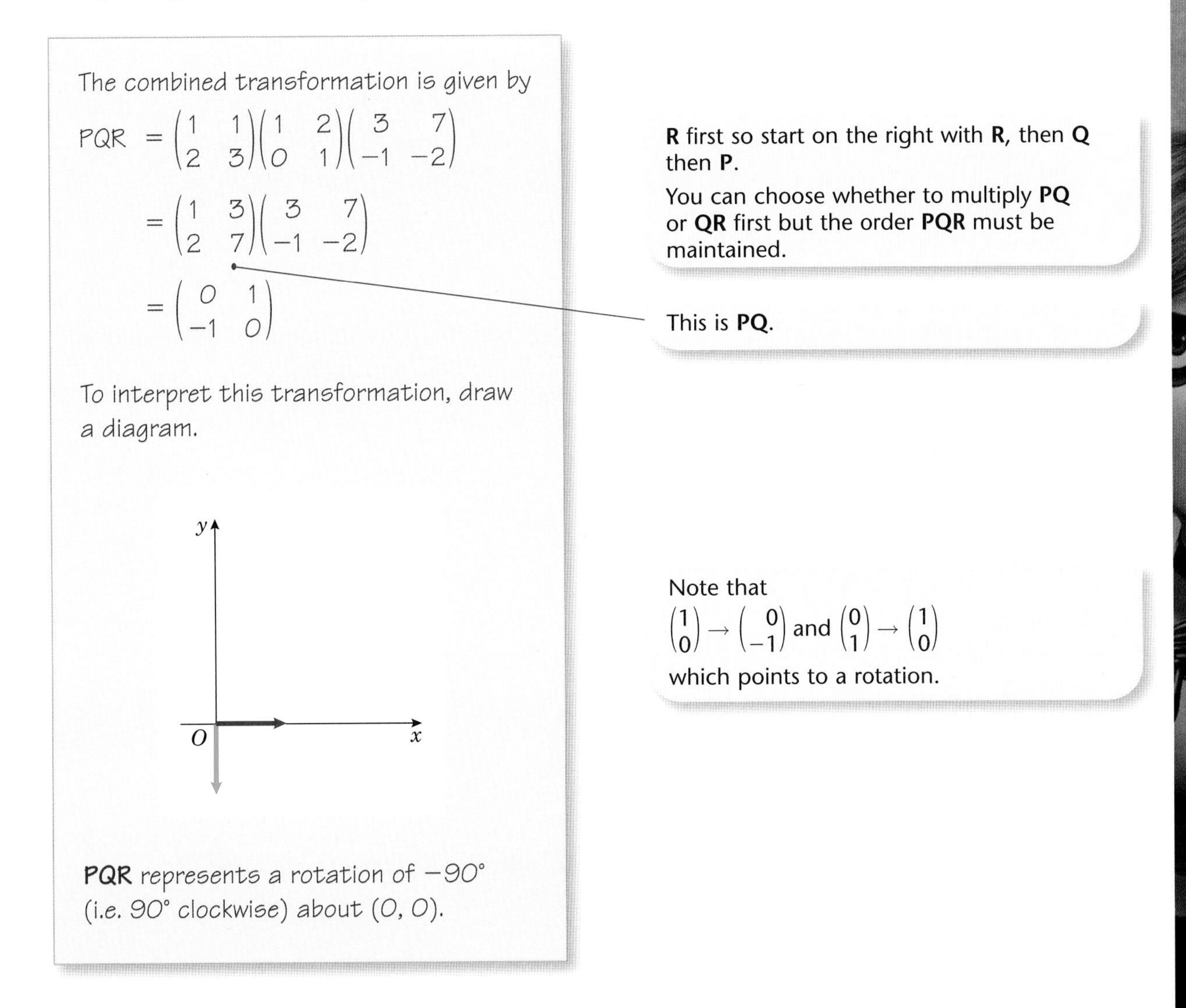

The combined transformation is given by

$$PQR = \begin{pmatrix} 1 & 1 \\ 2 & 3 \end{pmatrix}\begin{pmatrix} 1 & 2 \\ 0 & 1 \end{pmatrix}\begin{pmatrix} 3 & 7 \\ -1 & -2 \end{pmatrix}$$

$$= \begin{pmatrix} 1 & 3 \\ 2 & 7 \end{pmatrix}\begin{pmatrix} 3 & 7 \\ -1 & -2 \end{pmatrix}$$

$$= \begin{pmatrix} 0 & 1 \\ -1 & 0 \end{pmatrix}$$

R first so start on the right with **R**, then **Q** then **P**.
You can choose whether to multiply **PQ** or **QR** first but the order **PQR** must be maintained.

This is **PQ**.

To interpret this transformation, draw a diagram.

Note that $\begin{pmatrix} 1 \\ 0 \end{pmatrix} \to \begin{pmatrix} 0 \\ -1 \end{pmatrix}$ and $\begin{pmatrix} 0 \\ 1 \end{pmatrix} \to \begin{pmatrix} 1 \\ 0 \end{pmatrix}$ which points to a rotation.

PQR represents a rotation of $-90°$ (i.e. $90°$ clockwise) about $(0, 0)$.

Exercise 4F

1 $\mathbf{A} = \begin{pmatrix} -1 & 0 \\ 0 & -1 \end{pmatrix}$, $\mathbf{B} = \begin{pmatrix} 0 & -1 \\ -1 & 0 \end{pmatrix}$, $\mathbf{C} = \begin{pmatrix} 2 & 0 \\ 0 & 2 \end{pmatrix}$

Find these matrix products and describe the single transformation represented by the product.

a **AB** **b** **BA** **c** **AC**

d $\mathbf{A}^2$ **e** $\mathbf{C}^2$

2 A = rotation of 90° anticlockwise about (0, 0) B = rotation of 180° about (0, 0)
C = reflection in the x-axis D = reflection in the y-axis

a Find matrix representations of each of the four transformations A, B, C and D.

b Use matrix products to identify the single geometric transformation represented by each of these combinations.

i Reflection in the x-axis followed by a rotation of 180°about (0, 0).

ii Rotation of 180° about (0, 0) followed by a reflection in the x-axis.

iii Reflection in the y-axis followed by reflection in the x-axis.

iv Reflection in the y-axis followed by rotation of 90° about (0, 0).

v Rotation of 180° about (0, 0) followed by a second rotation of 180° about (0, 0).

vi Reflection in the x-axis followed by rotation of 90° about (0, 0) followed by a reflection in the y-axis.

vii Reflection in the y-axis followed by rotation of 180° about (0, 0) followed by a reflection in the x-axis.

3 Use a matrix product to find the single geometric transformation represented by a rotation of 270° anticlockwise about (0, 0) followed by a refection in the x-axis.

4 Use matrices to show that a refection in the y-axis followed by a reflection in the line $y = -x$ is equivalent to a rotation of 90° anticlockwise about (0, 0).

5 The matrix $\mathbf{R}$ is given by $\begin{pmatrix} \frac{1}{\sqrt{2}} & -\frac{1}{\sqrt{2}} \\ \frac{1}{\sqrt{2}} & \frac{1}{\sqrt{2}} \end{pmatrix}$.

a Find $\mathbf{R}^2$.

b Describe the geometric transformation represented by $\mathbf{R}^2$.

c Hence describe the geometric transformation represented by $\mathbf{R}$.

d Write down $\mathbf{R}^8$.

6 $\mathbf{P} = \begin{pmatrix} -5 & 2 \\ 3 & -1 \end{pmatrix}$, $\mathbf{Q} = \begin{pmatrix} -1 & -2 \\ 3 & 5 \end{pmatrix}$

The transformation represented by the matrix $\mathbf{R}$ is the result of the transformation represented by the matrix $\mathbf{P}$ followed by the transformation represented by the matrix $\mathbf{Q}$.

a Find $\mathbf{R}$.

b Give a geometrical interpretation of the transformation represented by $\mathbf{R}$.

7 $\mathbf{A} = \begin{pmatrix} 5 & -7 \\ 7 & -10 \end{pmatrix}$, $\mathbf{B} = \begin{pmatrix} 4 & 3 \\ 3 & 2 \end{pmatrix}$, $\mathbf{C} = \begin{pmatrix} -2 & 1 \\ -1 & 1 \end{pmatrix}$

Matrices **A**, **B** and **C** represent three transformations. By combining the three transformations in the order **B**, followed by **A**, followed by **C** a single transformation is obtained.
Find a matrix representation of this transformation and interpret it geometrically.

8 $\mathbf{P} = \begin{pmatrix} 1 & -5 \\ 1 & -1 \end{pmatrix}$, $\mathbf{Q} = \begin{pmatrix} 2 & 4 \\ 0 & 1 \end{pmatrix}$, $\mathbf{R} = \begin{pmatrix} 3 & 1 \\ -2 & 2 \end{pmatrix}$

Matrices **P**, **Q** and **R** represent three transformations. By combining the three transformations in the order **R**, followed by **Q**, followed by **P** a single transformation is obtained.
Find a matrix representation of this transformation and interpret it geometrically.

4.8 You can find the inverse of a 2 × 2 matrix where it exists.

- If $\mathbf{A} = \begin{pmatrix} a & b \\ c & d \end{pmatrix}$ then $\mathbf{A}^{-1} = \frac{1}{ad - bc}\begin{pmatrix} d & -b \\ -c & a \end{pmatrix}$

  and then $\mathbf{AA}^{-1} = \mathbf{A}^{-1}\mathbf{A} = \begin{pmatrix} 1 & 0 \\ 0 & 1 \end{pmatrix} = \mathbf{I}$

The value of $ad - bc$ is called the **determinant** of **A** and written det(**A**).

- $\mathbf{A} = \begin{pmatrix} a & b \\ c & d \end{pmatrix}$, $\det(\mathbf{A}) = ad - bc$ so $\mathbf{A}^{-1} = \frac{1}{\det(\mathbf{A})}\begin{pmatrix} d & -b \\ -c & a \end{pmatrix}$

Notice that if $\det(\mathbf{A}) = 0$ you will not be able to find $\mathbf{A}^{-1}$ because $\frac{1}{\det(\mathbf{A})}$ is not defined. In such cases we say **A** is **singular**.

- **If det(A) = 0, then A is a singular matrix and $\mathbf{A}^{-1}$ cannot be found.**
  If det(A) ≠ 0, then A is a non-singular matrix and $\mathbf{A}^{-1}$ exists.

Example 18

$\mathbf{A} = \begin{pmatrix} 3 & 2 \\ -1 & 1 \end{pmatrix}$, $\mathbf{B} = \begin{pmatrix} 2 & 1 \\ 2 & 1 \end{pmatrix}$, $\mathbf{C} = \begin{pmatrix} 1 & 3 \\ 2 & 0 \end{pmatrix}$

For each matrix **A**, **B** and **C**, determine whether or not the matrix is singular.
If the matrix is non-singular, find its inverse.

$\mathbf{A} = \begin{pmatrix} 3 & 2 \\ -1 & 1 \end{pmatrix}$ ∴ $\det(\mathbf{A}) = 3 \times 1 - 2 \times (-1)$

Use the determinant formula with $a = 3$, $b = 2$, $c = -1$ and $d = 1$.

$\det(\mathbf{A}) = 5$

Since $5 \neq 0$, **A** is non-singular

So $\mathbf{A}^{-1} = \frac{1}{5}\begin{pmatrix} 1 & -2 \\ 1 & 3 \end{pmatrix}$ or $\mathbf{A}^{-1} = \begin{pmatrix} 0.2 & -0.4 \\ 0.2 & 0.6 \end{pmatrix}$

Swap a and d and change the signs of b and c.

$\mathbf{A}^{-1}$ can be left in either form.

$\mathbf{B} = \begin{pmatrix} 2 & 2 \\ 1 & 1 \end{pmatrix}$ so $\det(\mathbf{B}) = 2 \times 1 - 1 \times 2 = 0$

Remember if det(**B**) = 0 then **B** is singular.

So **B** is singular and $\mathbf{B}^{-1}$ can not be found

$\mathbf{C} = \begin{pmatrix} 1 & 3 \\ 2 & 0 \end{pmatrix}$ ∴ $\det(\mathbf{C}) = 1 \times 0 - 3 \times 2 = -6$

Note that a determinant can be < 0

This is non-zero and so **C** is a non-singular matrix.

$\mathbf{C}^{-1} = -\frac{1}{6}\begin{pmatrix} 0 & -3 \\ -2 & 1 \end{pmatrix}$ or $\begin{pmatrix} 0 & \frac{1}{2} \\ \frac{1}{3} & -\frac{1}{6} \end{pmatrix}$

Swap a and d and change the signs of b and c. Then multiply by $\frac{1}{\det(\mathbf{C})}$.

Some calculators will find inverse matrices and you may wish to use your calculator to check your answer. You may be asked to find the inverse of a matrix in algebraic form.

Example 19

$\mathbf{A} = \begin{pmatrix} 4 & p+2 \\ -1 & 3-p \end{pmatrix}$

When $p = x$ the matrix $\mathbf{A}$ is singular.

a Find the value of x.

Given that $p \neq x$

b find $\mathbf{A}^{-1}$ leaving your answer in terms of p.

a $\det(A) = 4(3 - p) - (p + 2)(-1)$

$\det(A) = 12 - 4p + p + 2 = 14 - 3p$

A is singular so $\det(A) = 0$

$14 - 3x = 0$

$\Rightarrow \quad x = \frac{14}{3}$

Write $\mathbf{A} = \begin{pmatrix} a & b \\ c & d \end{pmatrix}$ and use the formula.

b $A^{-1} = \frac{1}{14 - 3p}\begin{pmatrix} 3-p & -(p+2) \\ 1 & 4 \end{pmatrix}$

Swap a and d and change the signs of b and c.

Example 20

P and **Q** are non-singular matrices. Prove that $(\mathbf{PQ})^{-1} = \mathbf{Q}^{-1}\mathbf{P}^{-1}$

Let $C = (PQ)^{-1}$ then $(PQ)C = I$

Use the definition of inverse $\mathbf{A}^{-1}\mathbf{A} = \mathbf{I} = \mathbf{A}\,\mathbf{A}^{-1}$.

$P^{-1}PQC = P^{-1}I$

Multiply on the left by $\mathbf{P}^{-1}$.

$(P^{-1}P)QC = P^{-1}I$

So $QC = P^{-1}$

Remember $\mathbf{P}^{-1}\mathbf{P} = \mathbf{I}$, $\mathbf{IQ} = \mathbf{Q}$ and $\mathbf{P}^{-1}\mathbf{I} = \mathbf{P}^{-1}$.

$Q^{-1}QC = Q^{-1}P^{-1}$

Multiply on the left by $\mathbf{Q}^{-1}$.

So $IC = Q^{-1}P^{-1}$

So $C = Q^{-1}P^{-1}$

Using $\mathbf{Q}^{-1}\mathbf{Q} = \mathbf{I}$.

■ **If A and B are non singular matrices then $(\mathbf{AB})^{-1} = \mathbf{B}^{-1}\mathbf{A}^{-1}$.**

Example 21

A and **B** are 2×2 non-singular matrices such that **BAB** = **I**.

a Prove that $\mathbf{A} = \mathbf{B}^{-1}\mathbf{B}^{-1}$

Given that $\mathbf{B} = \begin{pmatrix} 2 & 5 \\ 1 & 3 \end{pmatrix}$

b find the matrix **A** such that **BAB** = **I**.

a $\quad BAB = I$

$B^{-1}BAB = B^{-1}I$

i.e. $(B^{-1}B)AB = B^{-1}I$

so $AB = B^{-1}$

$AB\,B^{-1} = B^{-1}\,B^{-1}$

$AI = B^{-1}\,B^{-1}$

And hence $A = B^{-1}\,B^{-1}$

Multiply on the left by $\mathbf{B}^{-1}$.

Remember $\mathbf{B}^{-1}\mathbf{B} = \mathbf{I}$ and $\mathbf{B}^{-1}\mathbf{I} = \mathbf{B}^{-1}$.

Multiply on the right by $\mathbf{B}^{-1}$ and remember $\mathbf{B}\,\mathbf{B}^{-1} = \mathbf{I}$.

b $\quad B = \begin{pmatrix} 2 & 5 \\ 1 & 3 \end{pmatrix}$ so $\det(B) = 2 \times 3 - 5 \times 1 = 1$

First find $\mathbf{B}^{-1}$.

So $B^{-1} = \frac{1}{1}\begin{pmatrix} 3 & -5 \\ -1 & 2 \end{pmatrix} = \begin{pmatrix} 3 & -5 \\ -1 & 2 \end{pmatrix}$

From part **a**

$A = B^{-1}B^{-1} = \begin{pmatrix} 3 & -5 \\ -1 & 2 \end{pmatrix}\begin{pmatrix} 3 & -5 \\ -1 & 2 \end{pmatrix}$

i.e. $A = \begin{pmatrix} 14 & -25 \\ -5 & 9 \end{pmatrix}$

Use the result from part **a** and matrix multiplication to find **A**.

Exercise 4G

1 Determine which of these matrices are singular and which are non-singular. For those that are non-singular find the inverse matrix.

a $\begin{pmatrix} 3 & -1 \\ -4 & 2 \end{pmatrix}$ **b** $\begin{pmatrix} 3 & 3 \\ -1 & -1 \end{pmatrix}$ **c** $\begin{pmatrix} 2 & 5 \\ 0 & 0 \end{pmatrix}$

d $\begin{pmatrix} 1 & 2 \\ 3 & 5 \end{pmatrix}$ **e** $\begin{pmatrix} 6 & 3 \\ 4 & 2 \end{pmatrix}$ **f** $\begin{pmatrix} 4 & 3 \\ 6 & 2 \end{pmatrix}$

2 Find the value of a for which these matrices are singular.

a $\begin{pmatrix} a & 1+a \\ 3 & 2 \end{pmatrix}$ **b** $\begin{pmatrix} 1+a & 3-a \\ a+2 & 1-a \end{pmatrix}$ **c** $\begin{pmatrix} 2+a & 1-a \\ 1-a & a \end{pmatrix}$

3 Find inverses of these matrices.

a $\begin{pmatrix} a & 1+a \\ 1+a & 2+a \end{pmatrix}$ **b** $\begin{pmatrix} 2a & 3b \\ -a & -b \end{pmatrix}$

4 **a** Given that $\mathbf{ABC} = \mathbf{I}$, prove that $\mathbf{B}^{-1} = \mathbf{CA}$.

b Given that $\mathbf{A} = \begin{pmatrix} 0 & 1 \\ -1 & -6 \end{pmatrix}$ and $\mathbf{C} = \begin{pmatrix} 2 & 1 \\ -3 & -1 \end{pmatrix}$, find $\mathbf{B}$.

5 **a** Given that $\mathbf{AB} = \mathbf{C}$, find an expression for $\mathbf{B}$.

b Given further that $\mathbf{A} = \begin{pmatrix} 2 & -1 \\ 4 & 3 \end{pmatrix}$ and $\mathbf{C} = \begin{pmatrix} 3 & 6 \\ 1 & 22 \end{pmatrix}$, find $\mathbf{B}$.

6 **a** Given that $\mathbf{BAC} = \mathbf{B}$, where $\mathbf{B}$ is a non-singular matrix, find an expression for $\mathbf{A}$.

b When $\mathbf{C} = \begin{pmatrix} 5 & 3 \\ 3 & 2 \end{pmatrix}$, find $\mathbf{A}$.

7 The matrix $\mathbf{A} = \begin{pmatrix} 2 & -1 \\ -4 & 3 \end{pmatrix}$ and $\mathbf{AB} = \begin{pmatrix} 4 & 7 & -8 \\ -8 & -13 & 18 \end{pmatrix}$. Find the matrix $\mathbf{B}$.

8 The matrix $\mathbf{B} = \begin{pmatrix} 5 & -4 \\ 2 & 1 \end{pmatrix}$ and $\mathbf{AB} = \begin{pmatrix} 11 & -1 \\ -8 & 9 \\ -2 & -1 \end{pmatrix}$. Find the matrix $\mathbf{A}$.

9 The matrix $\mathbf{A} = \begin{pmatrix} 3a & b \\ 4a & 2b \end{pmatrix}$, where a and b are non-zero constants.

a Find $\mathbf{A}^{-1}$.

The matrix $\mathbf{B} = \begin{pmatrix} -a & b \\ 3a & 2b \end{pmatrix}$ and the matrix $\mathbf{X}$ is given by $\mathbf{B} = \mathbf{XA}$.

b Find $\mathbf{X}$.

10 The matrix $\mathbf{A} = \begin{pmatrix} a & 2a \\ b & 2b \end{pmatrix}$ and the matrix $\mathbf{B} = \begin{pmatrix} 2b & -2a \\ -b & a \end{pmatrix}$.

a Find det($\mathbf{A}$) and det($\mathbf{B}$).

b Find $\mathbf{AB}$.

11 The non-singular matrices $\mathbf{A}$ and $\mathbf{B}$ are commutative (i.e. $\mathbf{AB} = \mathbf{BA}$) and $\mathbf{ABA} = \mathbf{B}$.

a Prove that $\mathbf{A}^2 = \mathbf{I}$.

Given that $\mathbf{A} = \begin{pmatrix} 0 & 1 \\ 1 & 0 \end{pmatrix}$, by considering a matrix $\mathbf{B}$ of the form $\begin{pmatrix} a & b \\ c & d \end{pmatrix}$

b show that $a = d$ and $b = c$.

4.9 You can use inverse matrices to reverse the effect of a linear transformation.

Example 22

The triangle T has vertices at A, B and C. The matrix $\mathbf{M} = \begin{pmatrix} 4 & -1 \\ 3 & 1 \end{pmatrix}$ transforms T to the triangle T' with vertices at (4, 3), (4, 10) and (−4, −3).

a Find the coordinates of the points A, B and C.

b Sketch triangles T and T'.

c Show that the area of T' = area of T multiplied by det(**M**).

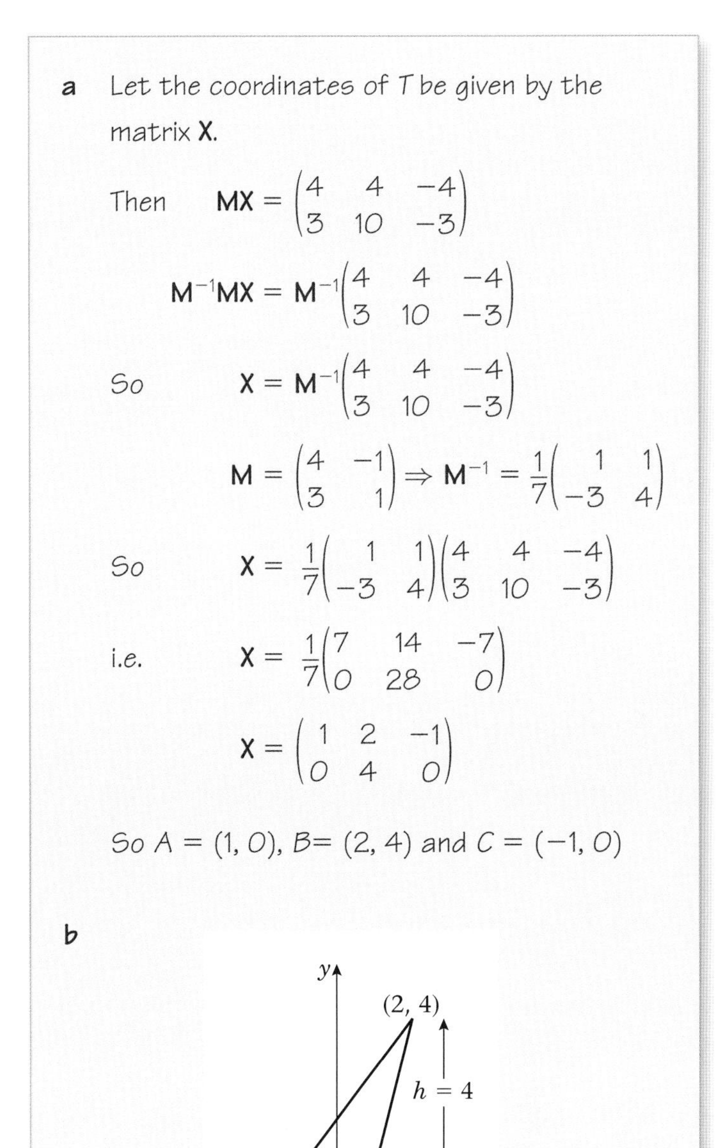

a Let the coordinates of T be given by the matrix **X**.

Then $\mathbf{MX} = \begin{pmatrix} 4 & 4 & -4 \\ 3 & 10 & -3 \end{pmatrix}$

$\mathbf{M}^{-1}\mathbf{MX} = \mathbf{M}^{-1}\begin{pmatrix} 4 & 4 & -4 \\ 3 & 10 & -3 \end{pmatrix}$

So $\mathbf{X} = \mathbf{M}^{-1}\begin{pmatrix} 4 & 4 & -4 \\ 3 & 10 & -3 \end{pmatrix}$

$\mathbf{M} = \begin{pmatrix} 4 & -1 \\ 3 & 1 \end{pmatrix} \Rightarrow \mathbf{M}^{-1} = \frac{1}{7}\begin{pmatrix} 1 & 1 \\ -3 & 4 \end{pmatrix}$

So $\mathbf{X} = \frac{1}{7}\begin{pmatrix} 1 & 1 \\ -3 & 4 \end{pmatrix}\begin{pmatrix} 4 & 4 & -4 \\ 3 & 10 & -3 \end{pmatrix}$

i.e. $\mathbf{X} = \frac{1}{7}\begin{pmatrix} 7 & 14 & -7 \\ 0 & 28 & 0 \end{pmatrix}$

$\mathbf{X} = \begin{pmatrix} 1 & 2 & -1 \\ 0 & 4 & 0 \end{pmatrix}$

So $A = (1, 0)$, $B = (2, 4)$ and $C = (-1, 0)$

b

Triangle T

Write the coordinates of T' as a 2×3 matrix $\begin{pmatrix} x_1 & x_2 & x_3 \\ y_1 & y_2 & y_3 \end{pmatrix}$.
Then **X** is a 2×3 matrix too.

Multiply on the left by $\mathbf{M}^{-1}$ and remember that $\mathbf{M}^{-1}\mathbf{M} = \mathbf{I}$.

Use the usual rules to find $\mathbf{M}^{-1}$.

It is usually simpler to keep the fraction ($\frac{1}{7}$ in this case) outside and multiply through with it in the final step.

Write down the coordinates of A, B and C from the columns of **X**.

A sketch need not be accurate but it should show the relative positions of the points and have some points marked to give a sense of scale.

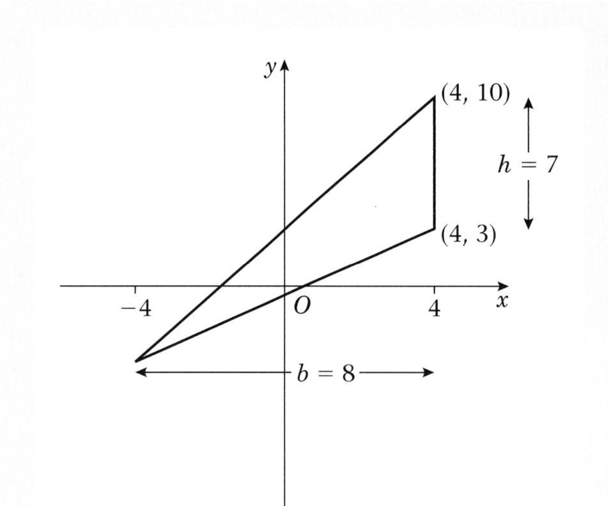

Triangle T'

c Area of $T = \frac{1}{2} \times 2 \times 4 = 4$

Area of $T' = \frac{1}{2} \times 8 \times 7 = 28$

$\det(\mathbf{M}) = 1 \times 4 - 3 \times (-1) = 7$

So area of T' = area of $T \times \det(\mathbf{M})$

since $28 = 4 \times 7$

Use $\frac{1}{2}bh$ with $b = 2$ and $h = 4$.

Use $\frac{1}{2}bh$ with $b = 8$ and $h = 7$ (the distance from (4, 3) to (4, 10)).

Exercise 4H

1 The matrix $\mathbf{R} = \begin{pmatrix} 0 & -1 \\ 1 & 0 \end{pmatrix}$

a Give a geometrical interpretation of the transformation represented by **R**.

b Find $\mathbf{R}^{-1}$.

c Give a geometrical interpretation of the transformation represented by $\mathbf{R}^{-1}$.

2 **a** The matrix $\mathbf{S} = \begin{pmatrix} -1 & 0 \\ 0 & -1 \end{pmatrix}$

i Give a geometrical interpretation of the transformation represented by **S**.

ii Show that $\mathbf{S}^2 = \mathbf{I}$.

iii Give a geometrical interpretation of the transformation represented by $\mathbf{S}^{-1}$.

b The matrix $\mathbf{T} = \begin{pmatrix} 0 & -1 \\ -1 & 0 \end{pmatrix}$

i Give a geometrical interpretation of the transformation represented by **T**.

ii Show that $\mathbf{T}^2 = \mathbf{I}$.

iii Give a geometrical interpretation of the transformation represented by $\mathbf{T}^{-1}$.

c Calculate det(**S**) and det(**T**) and comment on their values in the light of the transformations they represent.

3 The matrix $\mathbf{A}$ represents a reflection in the line $y = x$ and the matrix $\mathbf{B}$ represents an anticlockwise rotation of 270° about (0, 0).

a Find the matrix $\mathbf{C} = \mathbf{BA}$ and interpret it geometrically.

b Find $\mathbf{C}^{-1}$ and give a geometrical interpretation of the transformation represented by $\mathbf{C}^{-1}$.

c Find the matrix $\mathbf{D} = \mathbf{AB}$ and interpret it geometrically.

d Find $\mathbf{D}^{-1}$ and give a geometrical interpretation of the transformation represented by $\mathbf{D}^{-1}$.

4.10 You can use the determinant of a matrix to determine the area scale factor of the transformation.

- **As you will have seen in Example 22, for a transformation represented by a matrix M.**

Area of image = Area of object × |det(M)|

Sometimes the determinant is negative. In this formula you just use the value of the determinant. That is what $|\det(\mathbf{M})|$ means.

Example 23

The triangle OAB where A is (2, 0) and B is (1, 3) is transformed by the matrix $\mathbf{M} = \begin{pmatrix} p & -2 \\ 3 & 2 \end{pmatrix}$, where p is a positive constant.

a Find the coordinates of the images of A and B in terms of p.

b Given that the area of the image of OAB is 30, find the value of p.

c Sketch OAB and its image for the case $p = -3$ and comment.

a

$$\begin{pmatrix} p & -2 \\ 3 & 2 \end{pmatrix}\begin{pmatrix} 2 & 1 \\ 0 & 3 \end{pmatrix} \quad (\text{columns } A, B)$$

$$= \begin{pmatrix} 2p & p-6 \\ 6 & 9 \end{pmatrix}$$

So the image of A is $(2p, 6)$ and the image of B is $(p - 6, 9)$

Since (0, 0) is always unchanged by a matrix transformation, you only need to consider A and B. Write their coordinates as column vectors in a 2 × 2 matrix.

The first column gives the image of A and the second the image of B.

b

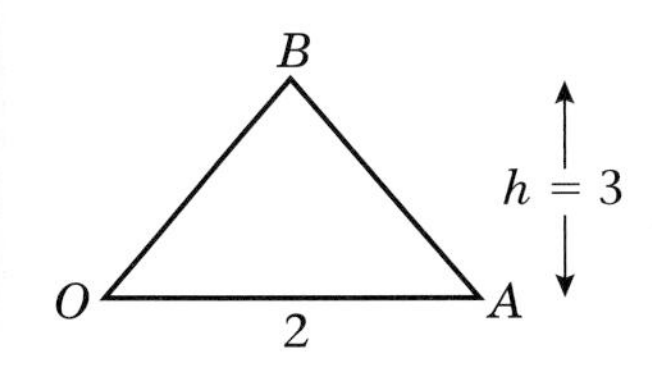

Sketch triangle OAB and use $\frac{1}{2}bh$.

So the area of $OAB = \frac{1}{2} \times 2 \times 3 = 3$

$\det(\mathbf{M}) = 2p - (-2) \times 3 = 2p + 6$

Area of image = area of object × $|\det(\mathbf{M})|$

Find $\det(\mathbf{M})$.

Use the determinant as an area scale factor.

$30 = 3(2p + 6)$
$10 = 2p + 6$
$4 = 2p$
So $p = 2$

If we had not been told that $p > 0$ then we could have $10 = -(2p + 6)$ leading to $p = -8$. Remember if $\det(\mathbf{M}) < 0$ this tells you that the transformation represented by **M** involves a reflection.

c When $p = -3$ then, using the answer to part **a**,
the image of A is $(-6, 6)$
the image of B is $(-9, 9)$

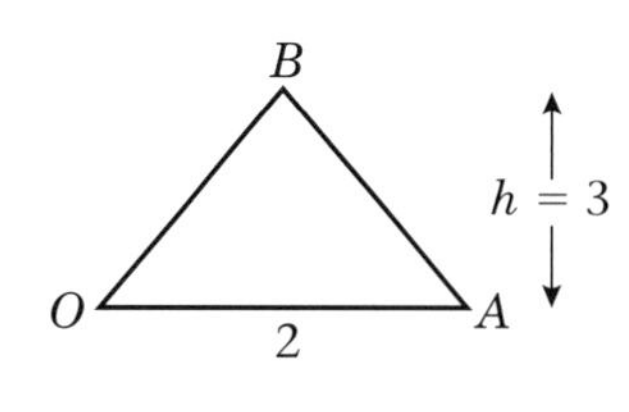

Sketch OAB and its image.

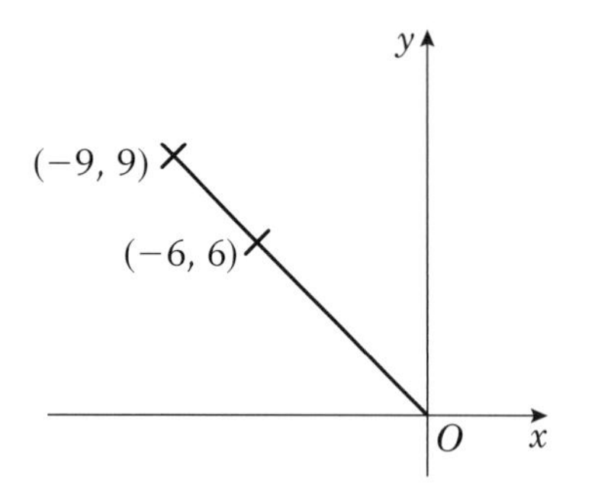

A singular matrix will always transform points onto a straight line.

The image of OAB is a straight line with equation $y = -x$.
$\det(\text{M}) = (-3) \times 2 - 3 \times (-2) = -6 + 6 = 0$
the transformation has 'squashed' the triangle into a straight line of zero area.

This is confirmed by considering the determinant which = 0 for singular matrices.

Exercise 4I

1 The matrix $\mathbf{A} = \begin{pmatrix} 2 & -1 \\ 4 & 3 \end{pmatrix}$ is used to transform the rectangle R with vertices at the points $(0, 0)$, $(0, 1)$, $(4, 1)$ and $(4, 0)$.

a Find the coordinates of the vertices of the image of R.

b Calculate the area of the image of R.

2 The triangle T has vertices at the points $(-3.5, 2.5)$, $(-16, 10)$ and $(-7, 4)$.

a Find the coordinates of the vertices of T under the transformation given by the matrix
$\mathbf{M} = \begin{pmatrix} -1 & -1 \\ 3 & 5 \end{pmatrix}$.

b Show that the area of the image of T is 7.5.

c Hence find the area of T.

3 The parallelogram P has vertices at the points $(-1, 0)$, $(0, -3)$, $(4, 0)$ and $(3, 3)$.

The matrix $\mathbf{A} = \begin{pmatrix} -2 & 3-a \\ 1 & a \end{pmatrix}$, where a is a constant.

a Find, in terms of a, the coordinates of the vertices of the image of P under the transformation given by $\mathbf{A}$.

b Find det($\mathbf{A}$), leaving your answer in terms of a.

Given that the area of the image of P is 75

c find the positive value of a.

4 $\mathbf{P} = \begin{pmatrix} 2 & -4 \\ 3 & 1 \end{pmatrix}$, $\mathbf{Q} = \begin{pmatrix} 1 & 2 \\ -1 & 4 \end{pmatrix}$, $\mathbf{R} = \begin{pmatrix} 1 & 2 \\ 2 & 1 \end{pmatrix}$.

A rectangle of area 5 cm^2 is transformed by the matrix $\mathbf{X}$ .Find the area of the image of the rectangle when $\mathbf{X}$ is:

a $\mathbf{P}$ **b** $\mathbf{Q}$ **c** $\mathbf{R}$

d $\mathbf{RQ}$ **e** $\mathbf{QR}$ **f** $\mathbf{RP}$

5 The triangle T has area 6 cm^2 and is transformed by the matrix $\begin{pmatrix} a & 3 \\ 3 & a+2 \end{pmatrix}$, where a is a constant, into triangle T'.

a Find det($\mathbf{A}$) in terms of a.

Given that the area of T' is 36 cm^2

b find the possible values of a.

4.11 You can use matrices and their inverses to solve linear simultaneous equations.

Example 24

Use an inverse matrix to solve the simultaneous equations
$2x - 3y = 5$, $-5x + 6y = -8$.

$$\begin{pmatrix} 2 & -3 \\ -5 & 6 \end{pmatrix}\begin{pmatrix} x \\ y \end{pmatrix} = \begin{pmatrix} 5 \\ -8 \end{pmatrix}$$

Write these equations as a matrix product. The LHS is a matrix **M**, made up of the coefficients of x and y from the equations, and a vector.

$$\mathbf{M} = \begin{pmatrix} 2 & -3 \\ -5 & 6 \end{pmatrix} \Rightarrow \mathbf{M}^{-1} = \frac{1}{-3}\begin{pmatrix} 6 & 3 \\ 5 & 2 \end{pmatrix}$$

Find the inverse of **M**.

$$\mathbf{M}\begin{pmatrix} x \\ y \end{pmatrix} = \begin{pmatrix} 5 \\ -8 \end{pmatrix}$$

$$\Rightarrow \mathbf{M}^{-1}\mathbf{M}\begin{pmatrix} x \\ y \end{pmatrix} = \mathbf{M}^{-1}\begin{pmatrix} 5 \\ -8 \end{pmatrix}$$

Multiply on the left by $\mathbf{M}^{-1}$.

$$\text{So } \begin{pmatrix} x \\ y \end{pmatrix} = \mathbf{M}^{-1}\begin{pmatrix} 5 \\ -8 \end{pmatrix} = -\frac{1}{3}\begin{pmatrix} 6 & 3 \\ 5 & 2 \end{pmatrix}\begin{pmatrix} 5 \\ -8 \end{pmatrix}$$

Since $\mathbf{M}^{-1}\mathbf{M} = \mathbf{I}$.

$$\begin{pmatrix} x \\ y \end{pmatrix} = -\frac{1}{3}\begin{pmatrix} 6 \\ 9 \end{pmatrix}$$

It is usually easier to leave a fraction such as $-\frac{1}{3}$ outside until the last step of the calculation.

$$\begin{pmatrix} x \\ y \end{pmatrix} = \begin{pmatrix} -2 \\ -3 \end{pmatrix}$$

So $x = -2$ and $y = -3$

Exercise 4J

1 Use inverse matrices to solve the following simultaneous equations

a $7x + 3y = 6$
$-5x - 2y = -5$

b $4x - y = -1$
$-2x + 3y = 8$

2 Use inverse matrices to solve the following simultaneous equations

a $4x - y = 11$
$3x + 2y = 0$

b $5x + 2y = 3$
$3x + 4y = 13$

Mixed exercise 4K

1 The matrix $\mathbf{A} = \begin{pmatrix} 3 & 1 \\ 4 & 2 \end{pmatrix}$ transforms the triangle PQR into the triangle with coordinates $(6, -2)$, $(4, 4)$, $(0, 8)$.

Find the coordinates of P, Q and R.

2 The matrix $\mathbf{A} = \begin{pmatrix} 1 & -3 \\ 2 & 1 \end{pmatrix}$ and $\mathbf{AB} = \begin{pmatrix} 4 & 1 & 9 \\ 1 & 9 & 4 \end{pmatrix}$.

Find the matrix **B**.

3 $\mathbf{A} = \begin{pmatrix} -2 & 1 \\ 7 & -3 \end{pmatrix}$, $\mathbf{B} = \begin{pmatrix} 4 & 1 \\ -5 & -1 \end{pmatrix}$, $\mathbf{C} = \begin{pmatrix} 3 & 1 \\ 2 & 1 \end{pmatrix}$.

The matrices **A**, **B** and **C** represent three transformations. By combining the three transformations in the order **A**, followed by **B**, followed by **C**, a simple single transformation is obtained which is represented by the matrix **R**.

a Find **R**.

b Give a geometrical interpretation of the transformation represented by **R**.

c Write down the matrix $\mathbf{R}^2$.

4 The matrix **Y** represents an anticlockwise rotation of 90° about (0, 0).

a Find **Y**.

The matrices **A** and **B** are such that **AB** = **Y**. Given that $\mathbf{B} = \begin{pmatrix} 3 & 2 \\ 2 & 1 \end{pmatrix}$

b find **A**.

c Simplify **ABABABAB**.

5 The matrix **R** represents a reflection in the x-axis and the matrix **E** represents an enlargement of scale factor 2 centre (0, 0).

a Find the matrix **C** = **ER** and interpret it geometrically.

b Find $\mathbf{C}^{-1}$ and give a geometrical interpretation of the transformation represented by $\mathbf{C}^{-1}$.

6 The quadrilateral R of area 4 cm² is transformed to R' by the matrix $\mathbf{P} = \begin{pmatrix} 1+p & p \\ 2-p & p \end{pmatrix}$, where p is a constant.

a Find det(**P**) in terms of p.

Given that the area of R'= 12 cm²

b find the possible values of p.

7 The matrix $\mathbf{A} = \begin{pmatrix} a & b \\ 2a & 3b \end{pmatrix}$, where a and b are non-zero constants.

a Find $\mathbf{A}^{-1}$.

The matrix $\mathbf{Y} = \begin{pmatrix} a & 2b \\ 2a & b \end{pmatrix}$ and the matrix **X** is given by **XA** = **Y**.

b Find **X**.

8 The 2 × 2, non-singular matrices **A**, **B** and **X** satisfy **XB** = **BA**.

a Find an expression for **X**.

b Given that $\mathbf{A} = \begin{pmatrix} 5 & 3 \\ 0 & -2 \end{pmatrix}$ and $\mathbf{B} = \begin{pmatrix} 2 & 1 \\ -1 & -1 \end{pmatrix}$, find **X**.

Summary of key points

1 An $n \times m$ matrix has n rows and m columns.

2 The transformation represented by the matrix product **ABC** means
first do the transformation represented by **C**
second do the transformation represented by **B**
third do the transformation represented by **A**.

3 The matrix $\mathbf{I} = \begin{pmatrix} 1 & 0 \\ 0 & 1 \end{pmatrix}$ is the **identity** matrix or transformation. It does not change a matrix or an object.

4 The **determinant** of a matrix $\mathbf{A} = \begin{pmatrix} a & b \\ c & d \end{pmatrix}$ is $\det(\mathbf{A}) = ad - bc$.

5 The **inverse** of a matrix $\mathbf{A} = \begin{pmatrix} a & b \\ c & d \end{pmatrix}$ is $\mathbf{A}^{-1} = \frac{1}{\det(\mathbf{A})}\begin{pmatrix} d & -b \\ -c & a \end{pmatrix}$.

After completing this chapter you will know how to:

- use the result for the sum of the first n natural numbers, $\sum_{r=1}^{n} r$
- use the results for the sum of the squares, and the sum of the cubes, of the first n natural numbers, $\sum_{r=1}^{n} r^2$ and $\sum_{r=1}^{n} r^3$ respectively
- use the results for $\sum_{r=1}^{n} 1$, $\sum_{r=1}^{n} r$, $\sum_{r=1}^{n} r^2$ and $\sum_{r=1}^{n} r^3$ to sum series where the general term is a polynomial in r of degree at most 3, e.g. $\sum_{r=1}^{n} (2r^3 + r^2 - 3r + 6)$.

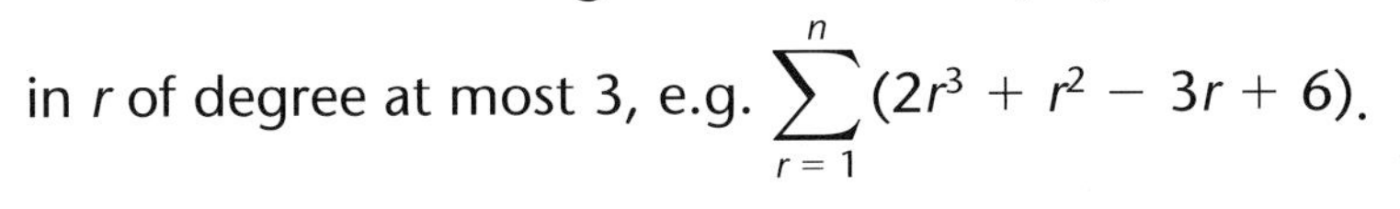

You will also be more familiar with the $\sum$ notation and know the result $\sum_{r=1}^{n} 1 = n$.

Series, and particularly infinite series, form an important part of the study of mathematics. Functions that you may be familiar with, such as $\sin x$, $\ln x$, e^x can all be written as infinite series: for example,

$e^x = \sum_{r=0}^{\infty} = 1 + x + \frac{x^2}{2!} + \ldots$. The fact that a wide range of functions can be expressed algebraically, as infinite series, is both a unifying aspect of the subject and a very useful one.

This chapter primarily considers a special group of finite series, the sums of powers of the first n natural numbers, $\sum_{r=1}^{n} r^k$, for $k = 0, 1, 2$ and 3.

Leonhard Euler introduced the notation '$\sum$' to mean the sum of a series.

5.1 You can use the $\sum$ notation.

- The sigma notation is a very useful, and concise way, to define a series. It makes further study of series more manageable.

- As you saw in Book C1, Chapter 6, where the $\sum$ notation was first introduced,

  $\sum_{r=1}^{n} U_r = U_1 + U_2 + U_3 + \dots U_n$, where U_r is a function of r.

- It is also used to mean 'the sum of the series'. For example, $\sum_{r=1}^{n} r^2 = 1^2 + 2^2 + 3^2 + \dots + n^2$, but you will also see that $\sum_{r=1}^{n} r^2 = \frac{n}{6}(n+1)(2n+1)$, the sum of the series (this result will be proved in Chapter 6).

- Series such as arithmetic series, geometric series and the binomial series, which you have already studied, can all be written in $\sum$ notation: for example, $\sum_{k=0}^{n-1} ar^k$ 'sums up' the geometric series $a + ar + ar^2 + \dots + ar^{n-1}$.

Example 1

Write out the series defined by the following

a $\sum_{r=1}^{n} (2r-1)$

b $\sum_{r=0}^{n-1} (2r+1)$

These expressions represent a general term in the series.

c $\sum_{k=1}^{n} (k^2+2)$

---

a $\sum_{r=1}^{n} (2r-1) = (2 \times 1 - 1) + (2 \times 2 - 1) + (2 \times 3 - 1) + (2 \times 4 - 1)$
$+ \dots + (2n-1) = 1 + 3 + 5 + 7 + \dots + (2n-1)$

b $\sum_{r=0}^{n-1} (2r+1) = (2 \times 0 + 1) + (2 \times 1 + 1) + (2 \times 2 + 1)$
$+ (2 \times 3 + 1) + \dots + [2(n-1) + 1] = 1 + 3 + 5 + 7 + \dots + (2n-1)$

The same series can be expressed in different ways in $\sum$ notation.

c $\sum_{k=1}^{n} (k^2+2) = (1^2+2) + (2^2+2) + (3^2+2) + (4^2+2) + \dots + (n^2+2)$
$= 3 + 6 + 11 + 18 + \dots + (n^2+2)$

Example 2

Write these series using the $\sum$ notation.

a $3 + 5 + 7 + 9 + 11 + 13$

b $2 + 5 + 10 + \ldots + (n^2 + 1)$

c $1 \times 2 + 2 \times 3 + 3 \times 4 + \ldots + (n - 2)(n - 1)$

a $3 + 5 + 7 + 9 + 11 + 13$ is the sum of the odd numbers from 3 to 13.
An odd number may be represented by $(2r - 1)$, where r is an integer.
The values of r corresponding to 3 and 13 are 2 and 7 respectively, so the series can be written as $\sum_{r=2}^{7}(2r - 1)$.

Equally $\sum_{r=1}^{6}(2r + 1)$ could be used.

b $2 + 5 + 10 + \ldots + (n^2 + 1) = \sum_{r=1}^{n}(r^2 + 1)$

c The general term of the series
$1 \times 2 + 2 \times 3 + 3 \times 4 + \ldots + (n - 2)(n - 1)$ can be written as $r(r + 1)$. Using this expression the first term corresponds to $r = 1$ and the final term corresponds to $r = (n - 2)$, so we can write $\sum_{r=1}^{n-2} r(r + 1)$.

Exercise 5A

1 Write out each of the following as a sum of terms, and hence calculate the sum of the series.

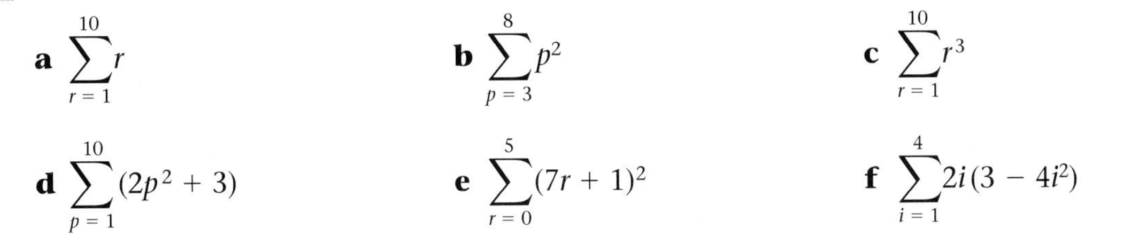

a $\sum_{r=1}^{10} r$ **b** $\sum_{p=3}^{8} p^2$ **c** $\sum_{r=1}^{10} r^3$

d $\sum_{p=1}^{10}(2p^2 + 3)$ **e** $\sum_{r=0}^{5}(7r + 1)^2$ **f** $\sum_{i=1}^{4} 2i(3 - 4i^2)$

2 Write each of the following as a sum of terms, showing the first three terms and the last term.

a $\sum_{r=1}^{n}(7r - 1)$ **b** $\sum_{r=1}^{n}(2r^3 + 1)$

c $\sum_{j=1}^{n}(j - 4)(j + 4)$ **d** $\sum_{p=3}^{k} p(p + 3)$

3 In each part of this question write out, as a sum of terms, the two series defined by $\sum f(r)$; for example, in part **c**, write out the series $\sum_{r=1}^{10} r^2$ and $\sum_{r=1}^{10} r$. Hence, state whether the given statements relating their sums are true or not.

a $\sum_{r=1}^{n}(3r+1) = \sum_{r=2}^{n+1}(3r-2)$ **b** $\sum_{r=1}^{n} 2r = \sum_{r=0}^{n} 2r$ **c** $\sum_{r=1}^{10} r^2 = \left(\sum_{1}^{10} r\right)^2$

d $\sum_{r=1}^{4} r^3 = \left(\sum_{r=1}^{4} r\right)^2$ **e** $\sum_{r=1}^{n}(3r^2+4) = 3\sum_{r=1}^{n} r^2 + 4$

4 Express these series using $\sum$ notation.

a $3 + 4 + 5 + 6 + 7 + 8 + 9 + 10$

b $1 + 8 + 27 + 64 + 125 + 216 + 343 + 512$

c $11 + 21 + 35 + \ldots + (2n^2 + 3)$

d $11 + 21 + 35 + \ldots + (2n^2 - 4n + 5)$

e $3 \times 5 + 5 \times 7 + 7 \times 9 + \ldots + (2r-1)(2r+1) + \ldots$ to k terms.

5.2 You can use the formula for the sum of the first *n* natural numbers, $\sum_{r=1}^{n} r$.

- **The sum of the first *n* natural numbers, $1 + 2 + 3 + \ldots + n$, is an arithmetic series of *n* terms, with $a = 1$ and $l = n$.**
- **Using the formula $S = \frac{n}{2}(a + l)$ for the sum of an arithmetic series, $\sum_{r=1}^{n} r = \frac{n}{2}(n + 1)$.**

Example 3

Evaluate **a** $\sum_{r=1}^{50} r$ **b** $\sum_{r=21}^{50} r$

a $\sum_{r=1}^{50} r = \frac{50 \times 51}{2} = 1275$

Substitute $n = 50$ in $\sum_{r=1}^{n} r = \frac{n}{2}(n + 1)$

b $\sum_{r=21}^{50} r = \sum_{r=1}^{50} r - \sum_{r=1}^{20} r = 1275 - \frac{20 \times 21}{2}$

$= 1275 - 210$

$= 1065.$

Substitute $n = 20$ in $\frac{n}{2}(n + 1)$

Example 4

Show that $\sum_{k=5}^{2N-1} k = 2N^2 - N - 10,\ N \geqslant 3$

$$\sum_{k=5}^{2N-1} k = \sum_{k=1}^{2N-1} k - \sum_{k=1}^{4} k$$

$$= \frac{(2N-1)[(2N-1)+1]}{2} - \frac{4 \times 5}{2}$$

Substitute $n = 2N - 1$, $n = 4$ in $\frac{n}{2}(n + 1)$.

$$= 2N^2 - N - 10$$

In general:

$$\sum_{r=k}^{n} f(r) = \sum_{r=1}^{n} f(r) - \sum_{r=1}^{k-1} f(r).$$

A common mistake is to use

$$\sum_{r=k}^{n} f(r) = \sum_{r=1}^{n} f(r) - \sum_{r=1}^{k} f(r).$$

Exercise 5B

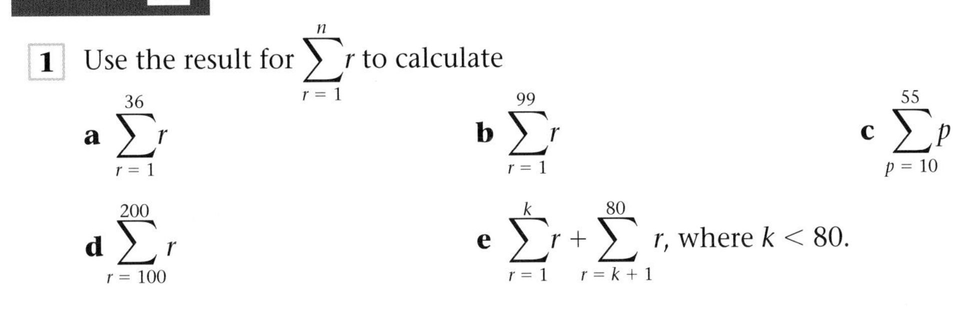

1 Use the result for $\sum_{r=1}^{n} r$ to calculate

a $\sum_{r=1}^{36} r$ **b** $\sum_{r=1}^{99} r$ **c** $\sum_{p=10}^{55} p$

d $\sum_{r=100}^{200} r$ **e** $\sum_{r=1}^{k} r + \sum_{r=k+1}^{80} r$, where $k < 80$.

2 Given that $\sum_{r=1}^{n} r = 528$,

a show that $n^2 + n - 1056 = 0$

b find the value of n.

3 **a** Find $\sum_{k=1}^{2n-1} k$.

b Hence show that $\sum_{k=n+1}^{2n-1} k = \frac{3n}{2}(n - 1),\ n \geqslant 2$.

4 Show that $\sum_{r=k-1}^{2k} r = \frac{(k+2)(3k-1)}{2},\ k \geqslant 1$

5 **a** Show that $\sum_{r=1}^{n^2} r - \sum_{r=1}^{n} r = \frac{n(n^3 - 1)}{2}$.

b Hence evaluate $\sum_{r=10}^{81} r$.

- **All series of the form $\sum_{r=1}^{n}(ar + b)$, where a and b are constants, are arithmetic series and so their sums can always be found by using the appropriate formula.**
- **However, by splitting up $\sum_{r=1}^{n}(ar + b)$ into $a\sum_{r=1}^{n} r + b\sum_{r=1}^{n} 1$, you can find such sums using the results for $\sum_{r=1}^{n} r$ and $\sum_{r=1}^{n} 1$.**

Example 5

Show that $\sum_{r=1}^{n}(3r + 2)$ can be written as $3\sum_{r=1}^{n} r + 2\sum_{1}^{n} 1$.

$$\sum_{r=1}^{n}(3r+2) = (3 \times 1 + 2) + (3 \times 2 + 2) + (3 \times 3 + 2) + \ldots + (3 \times n + 2)$$
$$= (3 \times 1 + 3 \times 2 + 3 \times 3 + \ldots + 3 \times n) + (2 + 2 + 2 + \ldots + 2)$$
$$= 3 \times (1 + 2 + 3 + \ldots + n) + 2\underbrace{(1 + 1 + 1 + \ldots + 1)}_{\text{sum of } n \text{ ones}}$$
$$= 3\sum_{r=1}^{n} r + 2\sum_{r=1}^{n} 1$$
$$\text{where } \sum_{r=1}^{n} 1 = \underbrace{(1 + 1 + 1 + \ldots + 1)}_{\text{sum of } n \text{ ones}} = n$$

As $r^0 = 1$ for all positive integers r, the sum of n ones, $(1 + 1 + 1 + \ldots + 1)$ can be written as $(1^0 + 2^0 + 3^0 + \ldots + n^0)$ $= \sum_{r=1}^{n} r^0 = \sum_{r=1}^{n} 1$.

Example 6

Evaluate $\sum_{r=1}^{25}(3r + 1)$

$$\sum_{1}^{25}(3r+1) = 3\sum_{1}^{25} r + \sum_{1}^{25} 1$$
$$= 3\frac{25 \times 26}{2} + 25$$
$$= 975 + 25 = 1000$$

Example 7

a Show that $\sum_{r=1}^{n}(7r - 4) = \frac{n}{2}(7n - 1)$.

b Hence calculate $\sum_{r=20}^{50}(7r - 4)$.

a $\sum_{r=1}^{n}(7r - 4) = 7\sum_{r=1}^{n} r - 4\sum_{r=1}^{n} 1 = 7\frac{n}{2}(n + 1) - 4n$

$$= \frac{7n^2 + 7n - 8n}{2}$$

$$= \frac{n}{2}(7n - 1)$$

It is a common mistake to write this as 4, not $4n$.

b $\sum_{r=20}^{50}(7r - 4) = \sum_{r=1}^{50}(7r - 4) - \sum_{r=1}^{19}(7r - 4)$

$$= \frac{50}{2}\{7(50) - 1\} - \frac{19}{2}\{7(19) - 1\}$$

$$= 8725 - 1254$$

$$= 7471$$

Using the result from **a**.

Exercise 5C

(In this exercise use the results for $\sum_{r=1}^{n} r$ and $\sum_{r=1}^{n} 1$.)

1 Calculate the sum of the series:

a $\sum_{r=1}^{55}(3r - 1)$ **b** $\sum_{r=1}^{90}(2 - 7r)$ **c** $\sum_{r=1}^{46}(9 + 2r)$

2 Show that

a $\sum_{r=1}^{n}(3r + 2) = \frac{n}{2}(3n + 7)$ **b** $\sum_{i=1}^{2n}(5i - 4) = n(10n - 3)$

c $\sum_{r=1}^{n+2}(2r + 3) = (n + 2)(n + 6)$ **d** $\sum_{p=3}^{n}(4p + 5) = (2n + 11)(n - 2)$

3 **a** Show that $\sum_{r=1}^{k}(4r - 5) = 2k^2 - 3k$.

b Find the smallest value of k for which $\sum_{r=1}^{k}(4r - 5) > 4850$.

4 Given that $u_r = ar + b$ and $\sum_{r=1}^{n} u_r = \frac{n}{2}(7n + 1)$, find the constants a and b.

5 **a** Show that $\sum_{r=1}^{4n-1} (1 + 3r) = 24n^2 - 2n - 1,\ n \geqslant 1$

b Hence calculate $\sum_{r=1}^{99} (1 + 3r)$.

6 Show that $\sum_{r=1}^{2k+1} (4 - 5r) = -(2k + 1)(5k + 1),\ k \geqslant 0$

5.3 You can use formulae for the sum of the squares of the first n natural numbers, $\sum_{r=1}^{n} r^2$, and for the sum of the cubes of the first n natural numbers $\sum_{r=1}^{n} r^3$.

The standard results

$$\sum_{r=1}^{n} \boldsymbol{r^2} = 1^2 + 2^2 + 3^2 + 4^2 + \ldots + n^2 = \boldsymbol{\frac{n}{6}(n + 1)(2n + 1)}$$

$$\sum_{r=1}^{n} \boldsymbol{r^3} = 1^3 + 2^3 + 3^3 + 4^3 + \ldots + n^3 = \boldsymbol{\frac{n^2}{4}(n + 1)^2}$$

will be used without proof in this Chapter. Proofs will be given in Chapter 6.

Example 8

Evaluate **a** $\sum_{r=20}^{40} r^2$ **b** $\sum_{r=1}^{25} r^3$

a $\sum_{r=20}^{40} r^2 = \sum_{r=1}^{40} r^2 - \sum_{r=1}^{19} r^2$

$= \frac{40}{6}(40 + 1)(80 + 1) - \frac{19}{6}(19 + 1)(38 + 1)$

$= 22\,140 - 2470 = 19\,670$

Substitute $n = 40$ and $n = 19$ in $\sum_{r=1}^{n} r^2 = \frac{n}{6}(n + 1)(2n + 1)$.

b $\sum_{r=1}^{25} r^3 = \frac{25^2 \times 26^2}{4} = 105\,625$

Substitute $n = 25$ in $\frac{n^2}{4}(n + 1)^2$.

Example 9

a Find $\sum_{r=n+1}^{2n} r^2$.

b Verify that the result is true for $n = 1$ and $n = 2$.

a $\sum_{r=n+1}^{2n} r^2 = \sum_{r=1}^{2n} r^2 - \sum_{r=1}^{n} r^2$

$= \frac{2n}{6}(2n+1)(4n+1) - \frac{n}{6}(n+1)(2n+1)$

Replacing n by $2n$ in $\frac{n}{6}(n+1)(2n+1)$.

$= \frac{n}{6}(2n+1)\{2(4n+1) - (n+1)\}$

$= \frac{n}{6}(2n+1)(7n+1)$

b When $n = 1$,

$\sum_{r=n+1}^{2n} r^2 = \sum_{r=2}^{2} r^2 = 2^2 = 4,\ \frac{n}{6}(2n+1)(7n+1) = \frac{1}{6}(3)(8) = 4$ ✓

When $n = 2$,

$\sum_{r=n+1}^{2n} r^2 = \sum_{r=3}^{4} r^2 = 3^2 + 4^2 = 25,\ \frac{n}{6}(2n+1)(7n+1) = \frac{2}{6}(5)(15) = 25$ ✓

When you have been asked to find a general result for a sum it is good practice to test it for small values of n. It will not prove that you are correct, but if one value of n does not work, you know that your result is incorrect.

Exercise 5D

1 Verify that $\sum_{r=1}^{n} r^2 = \frac{n}{6}(n+1)(2n+1)$ is true for $n = 1$, 2 and 3.

2 **a** By writing out each series, evaluate $\sum_{r=1}^{n} r$ for $n = 1$, 2, 3 and 4.

b By writing out each series, evaluate $\sum_{r=1}^{n} r^3$ for $n = 1$, 2, 3 and 4.

c What do you notice about the corresponding results for each value of n?

3 Using the appropriate formula, evaluate **a** $\sum_{r=1}^{100} r^2$ **b** $\sum_{r=20}^{40} r^2$ **c** $\sum_{r=1}^{30} r^3$ **d** $\sum_{r=25}^{45} r^3$

4 Use the formula for $\sum_{r=1}^{n} r^2$ or $\sum_{r=1}^{n} r^3$ to find the sum of

a $1^2 + 2^2 + 3^2 + 4^2 + \dots + 52^2$

b $2^3 + 3^3 + 4^3 + \dots + 40^3$

c $26^2 + 27^2 + 28^2 + 29^2 + \dots + 100^2$

d $1^2 + 2^2 + 3^2 + \dots + (k+1)^2$

e $1^3 + 2^3 + 3^3 + \dots + (2n-1)^3$

5 For each of the following series write down, in terms of n, the sum, giving the result in its simplest form

a $\sum_{r=1}^{2n} r^2$ **b** $\sum_{r=1}^{n^2-1} r^2$ **c** $\sum_{i=1}^{2n-1} i^2$

d $\sum_{r=1}^{n+1} r^3$ **e** $\sum_{k=n+1}^{3n} k^3,\ n > 0.$

6 Show that

a $\sum_{r=2}^{n} r^2 = \frac{1}{6}(n-1)(2n^2+5n+6)$ **b** $\sum_{r=n}^{2n} r^2 = \frac{n}{6}(n+1)(14n+1)$

7 **a** Show that $\sum_{k=n}^{2n} k^3 = \dfrac{3n^2(n+1)(5n+1)}{4}$

b Find $\sum_{k=30}^{60} k^3$.

8 **a** Show that $\sum_{r=1}^{2n} r^3 = n^2(2n+1)^2$.

b By writing out the series for $\sum_{r=1}^{n} (2r)^3$, show that $\sum_{r=1}^{n} (2r)^3 = 8\sum_{r=1}^{n} r^3$.

c Show that $1^3 + 3^3 + 5^3 + \ldots + (2n-1)^3$ can be written as $\sum_{r=1}^{2n} r^3 - \sum_{r=1}^{n} (2r)^3$.

d Hence show that the sum of the cubes of the first n odd natural numbers, $1^3 + 3^3 + 5^3 + \ldots + (2n-1)^3$, is $n^2(2n^2-1)$.

5.4 You can use known formulae to sum more complex series.

In Example 5 we saw that $\sum_{r=1}^{n} (3r+2)$ can be written as $3\sum_{r=1}^{n} r + 2\sum_{1}^{n} 1$. This demonstrates the use of two general rules relating to sigma notation.

- **addition rule $\sum_{r=1}^{n} (U_r + V_r) = \sum_{r=1}^{n} U_r + \sum_{r=1}^{n} V_r$**

- **multiple rule $\sum_{r=1}^{n} kU_r = k\sum_{r=1}^{n} U_r$, where k is a constant**

- **These rules mean that more complicated sums like $\sum_{r=1}^{n} (ar^2 + br + c)$ and $\sum_{r=1}^{n} (ar^3 + br^2 + cr + d)$ can be found using the results for $\sum_{r=1}^{n} 1$, $\sum_{r=1}^{n} r$, $\sum_{r=1}^{n} r^2$ and $\sum_{r=1}^{n} r^3$.**

Example 10

a Show that $\sum_{r=1}^{n}(r^2 + r - 2) = \frac{n}{3}(n + 4)(n - 1)$

b Deduce the sum of the series $4 + 10 + 18 + 28 + 40 + \ldots + 418$

a $\sum_{r=1}^{n}(r^2 + r - 2) = \sum_{r=1}^{n} r^2 + \sum_{r=1}^{n}(r - 2)$

Using the addition rule and the multiple rule, any number of terms can be 'split up'.

$= \sum_{r=1}^{n} r^2 + \sum_{r=1}^{n} r - 2\sum_{r=1}^{n} 1$

$= \frac{n}{6}(n + 1)(2n + 1) + \frac{n}{2}(n + 1) - 2n$

Use the results for $\sum_{r=1}^{n} r^2$, $\sum_{r=1}^{n} r$ and $\sum_{r=1}^{n} 1$

$= \frac{n}{6}[(n + 1)(2n + 1) + 3(n + 1) - 12]$

$= \frac{n}{6}[2n^2 + 3n + 1 + 3n + 3 - 12]$

$= \frac{n}{6}[2n^2 + 6n - 8]$

$= \frac{n}{3}[n^2 + 3n - 4]$

$= \frac{n}{3}(n + 4)(n - 1)$

b $\sum_{r=1}^{20}(r^2 + r - 2) = 0 + 4 + 10 + 18 + 28 + 40 + \ldots + 418$

So $4 + 10 + 18 + 28 + 40 + \ldots + 418$

$= \frac{20}{3}(20 + 4)(20 - 1) = 3040$

Substitute $n = 20$ in $\frac{n}{3}(n + 4)(n - 1)$.

Example 11

a Find the sum of the series $\sum_{r=1}^{n} r(r + 3)(2r - 1)$

b Hence calculate $\sum_{r=11}^{40} r(r + 3)(2r - 1)$

a $\sum_{r=1}^{n} r(r+3)(2r-1) = \sum_{r=1}^{n}(2r^3 + 5r^2 - 3r)$

First multiply out the brackets.

$= 2\sum_{r=1}^{n} r^3 + 5\sum_{r=1}^{n} r^2 - 3\sum_{r=1}^{n} r$

Use the addition and multiple rules.

$= \frac{2n^2}{4}(n+1)^2 + 5\frac{n}{6}(n+1)(2n+1) - 3\frac{n}{2}(n+1)$

Using the results for $\sum_{r=1}^{n} r^3, \sum_{r=1}^{n} r^2$ and $\sum_{r=1}^{n} r$.

$= \frac{n(n+1)}{6}[3n(n+1) + 5(2n+1) - 9]$

$= \frac{n(n+1)}{6}[3n^2 + 13n - 4]$

b $\sum_{r=11}^{40} r(r+3)(2r-1) = \sum_{r=1}^{40} r(r+3)(2r-1) - \sum_{r=1}^{10} r(r+3)(2r-1)$

$= \frac{40 \times 41 \times 5316}{6} - \frac{10 \times 11 \times 426}{6}$

Substitute $n = 40$ and $n = 10$ in the result for **a**.

$= 1453040 - 7810$

$= 1445230$

Exercise 5E

1 Use the formulae for $\sum_{r=1}^{n} r^3, \sum_{r=1}^{n} r^2, \sum_{r=1}^{n} r$ and $\sum_{r=1}^{n} 1$, where appropriate, to find

a $\sum_{m=1}^{30}(m^2 - 1)$ **b** $\sum_{r=1}^{40} r(r+4)$ **c** $\sum_{r=1}^{80} r(r^2+3)$ **d** $\sum_{r=11}^{35}(r^3 - 2)$.

2 Use the formulae for $\sum_{r=1}^{n} r^3, \sum_{r=1}^{n} r^2$, and $\sum_{r=1}^{n} r$, where appropriate, to find

a $\sum_{r=1}^{n}(r^2 + 4r)$ **b** $\sum_{r=1}^{n} r(2r^2 - 1)$ **c** $\sum_{r=1}^{2n} r^2(1+r)$,

giving your answer in its simplest form.

3 **a** Write out $\sum_{r=1}^{n} r(r+1)$ as a sum, showing at least the first three terms and the final term.

b Use the results for $\sum_{r=1}^{n} r$ and $\sum_{r=1}^{n} r^2$ to calculate

$1 \times 2 + 2 \times 3 + 3 \times 4 + 4 \times 5 + 5 \times 6 + \ldots + 60 \times 61$.

4 **a** Show that $\sum_{r=1}^{n}(r+2)(r+5)=\frac{n}{3}(n^2+12n+41)$.

b Hence calculate $\sum_{r=10}^{50}(r+2)(r+5)$.

5 **a** Show that $\sum_{r=2}^{n}(r-1)r(r+1)=\frac{(n-1)n(n+1)(n+2)}{4}$.

b Hence find the sum of the series
$13 \times 14 \times 15 + 14 \times 15 \times 16 + 15 \times 16 \times 17 + \ldots + 44 \times 45 \times 46$.

6 Find the following sums, and check your results for the cases $n = 1, 2$ and 3.

a $\sum_{r=1}^{n}(r^3-1)$ **b** $\sum_{r=1}^{n}(2r-1)^2$ **c** $\sum_{r=1}^{n}r(r+1)^2$

7 **a** Show that $\sum_{r=1}^{n}r^2(r-1)=\frac{n}{12}(n^2-1)(3n+2)$.

b Deduce the sum of $1 \times 2^2 + 2 \times 3^2 + 3 \times 4^2 + \ldots + 30 \times 31^2$.

8 **a** Show that $\sum_{r=2}^{n}(r-1)(r+1)=\frac{n}{6}(2n+5)(n-1)$.

b Hence sum the series $1 \times 3 + 2 \times 4 + 3 \times 5 + \ldots + 35 \times 37$.

9 **a** Write out the series defined by $\sum_{r=7}^{12}r(2+3r)$, and hence find its sum.

b Show that $\sum_{r=n+1}^{2n}r(2+3r)=\frac{n}{2}(14n^2+15n+3)$.

c By substituting the appropriate value of n into the formula in **b**, check that your answer for **a** is correct.

10 Find the sum of the series $1 \times 1 + 2 \times 3 + 3 \times 5 + \ldots$ to n terms.

Mixed exercise 5F

1 **a** Write down the first three terms and the last term of the series given by $\sum_{r=1}^{n}(2r+3^r)$.

b Find the sum of this series.

c Verify that your result in **b** is correct for the cases $n = 1, 2$ and 3.

2 Find **a** $\sum_{r=1}^{50}(7r+5)$ **b** $\sum_{r=1}^{40}(2r^2-1)$ **c** $\sum_{r=33}^{75}r^3$.

3 Given that $\sum_{r=1}^{n}U_r=n^2+4n$,

a find $\sum_{r=1}^{n-1}U_r$. **b** Deduce an expression for U_n. **c** Find $\sum_{r=n}^{2n}U_r$.

4 Evaluate $\sum_{r=1}^{30}r(3r-1)$ E

5 Find $\sum_{r=1}^{n} r^2(r-3)$.

6 Show that $\sum_{r=1}^{2n} (2r-1)^2 = \frac{2n}{3}(16n^2-1)$. **E**

7 **a** Show that $\sum_{r=1}^{n} r(r+2) = \frac{n}{6}(n+1)(2n+7)$. **E**

b Using this result , or otherwise, find in terms of n, the sum of

$3\log2 + 4\log2^2 + 5\log2^3 + \dots + (n+2)\log2^n$.

8 Show that $\sum_{r=n}^{2n} r^2 = \frac{n}{6}(n+1)(an+b)$, where a and b are constants to be found. **E**

9 **a** Show that $\sum_{r=1}^{n} (r^2-r-1) = \frac{n}{3}(n-2)(n+2)$. **E**

b Hence calculate $\sum_{r=10}^{40} (r^2-r-1)$.

10 **a** Show that $\sum_{r=1}^{n} r(2r^2+1) = \frac{n}{2}(n+1)(n^2+n+1)$.

b Hence calculate $\sum_{r=26}^{58} r(2r^2+1)$.

11 Find **a** $\sum_{r=1}^{n} r(3r-1)$ **b** $\sum_{r=1}^{n} (r+2)(3r+5)$ **c** $\sum_{r=1}^{n} (2r^3-2r+1)$.

12 **a** Show that $\sum_{r=1}^{n} r(r+1) = \frac{n}{3}(n+1)(n+2)$.

b Hence calculate $\sum_{r=31}^{60} r(r+1)$.

13 **a** Show that $\sum_{r=1}^{n} r(r+1)(r+2) = \frac{n}{4}(n+1)(n+2)(n+3)$.

b Hence evaluate $3 \times 4 \times 5 + 4 \times 5 \times 6 + 5 \times 6 \times 7 + \dots + 40 \times 41 \times 42$.

14 **a** Show that $\sum_{r=1}^{n} r\{2(n-r)+1\} = \frac{n}{6}(n+1)(2n+1)$.

b Hence sum the series $(2n-1) + 2(2n-3) + 3(2n-5) + \dots + n$

15 **a** Show that when n is even,

$$1^3 - 2^3 + 3^3 - \dots - n^3 = 1^3 + 2^3 + 3^3 + \dots + n^3 - 16\left(1^3 + 2^3 + 3^3 + \dots + \left(\frac{n}{2}\right)^3\right)$$

$$= \sum_{r=1}^{n} r^3 - 16\sum_{r=1}^{\frac{n}{2}} r^3.$$

b Hence show that, for n even, $1^3 - 2^3 + 3^3 - \dots - n^3 = -\frac{n^2}{4}(2n+3)$

c Deduce the sum of $1^3 - 2^3 + 3^3 - \dots - 40^3$.

Summary of key points

1 The notation $\sum_{r=1}^{n} U_r$ defines the series $U_1 + U_2 + U_3 + \ldots + U_n$, where U_r is a general term in the series; it also is used to mean 'sum the series'.

2 If the series is summed from $r = k$ to $r = n$, then $\sum_{r=k}^{n} U_r = \sum_{r=1}^{n} U_r - \sum_{r=1}^{k-1} U_r$.

3 The sums of powers, k, of the first n natural numbers, for the cases $k = 0, 1, 2$ and 3 are

- $\sum_{r=1}^{n} r^0 = \sum_{r=1}^{n} 1 = 1 + 1 + 1 \ldots + 1 = n$
- $\sum_{r=1}^{n} r = 1 + 2 + 3 + \ldots + n = \frac{n}{2}(n + 1)$
- $\sum_{r=1}^{n} r^2 = 1^2 + 2^2 + 3^2 + \ldots + n^2 = \frac{n}{6}(n + 1)(2n + 1)$
- $\sum_{r=1}^{n} r^3 = 1^3 + 2^3 + 3^3 + \ldots + n^3 = \frac{n^2}{4}(n + 1)^2$

These should be learnt.

Note:
$$\sum_{r=1}^{n} r^3 = \left(\sum_{r=1}^{n} r\right)^2$$

4 If the general term is a more complicated function of r, the sum can be 'split up' so that some, or all, of the above basic results can be used.

For example:

a $\sum_{r=1}^{n} (r^3 + 3r^2 - 2r + 5) = \sum_{r=1}^{n} r^3 + 3\sum_{r=1}^{n} r^2 - 2\sum_{r=1}^{n} r + 5\sum_{r=1}^{n} 1,$

b $\sum_{r=1}^{n} (2r + 1)(r + 2) = \sum_{r=1}^{n} (2r^2 + 5r + 2) = 2\sum_{r=1}^{n} r^2 + 5\sum_{r=1}^{n} r + 2\sum_{r=1}^{n} 1.$

After completing this chapter you will know how to:

- use the method of mathematical induction to prove general statements which involve positive integers.

Proof by mathematical induction

The structure of a proof by induction can be compared with a line of falling dominoes. If the first domino falls, then so will its neighbour, until all the dominoes fall.

The first example of rigour in proof by induction was a proof that the sum of the first n odd numbers is n^2. This was given by Francesco Maurolico (1494–1575).

6.1 You can obtain a proof for the summation of a series, using induction.

$n = 1$; $1 = 1$

$n = 2$; $1 + 3 = 4$

$n = 3$; $1 + 3 + 5 = 9$

$n = 4$; $1 + 3 + 5 + 7 = 16$

$n = 5$; $1 + 3 + 5 + 7 + 9 = 25$

$n = 6$; $1 + 3 + 5 + 7 + 9 + 11 = 36$

etc.

Looking at the summations above we can easily spot a general result which is true.

General Statement: If you add up the first n positive odd numbers you obtain the squared number n^2.

This general statement can be written as:

$$1 + 3 + 5 + 7 + 9 + \ldots + (2n - 1) = n^2$$

Using the series notation learned in Chapter 5 the general statement can be also rewritten as:

$$\sum_{r=1}^{n} (2r - 1) = n^2.$$

The diagram below is also a pictorial representation of the result for $n = 1, 2, 3, 4$ and $n = 5$.

$n = 1$ $n = 2$ $n = 3$ $n = 4$ $n = 5$

This may give more insight about why the **general statement** is true.

In this chapter, however, you will learn a more formal method of proof, known as **proof by induction**, to prove general statements, such as the one above, which apply to positive integer values, n, where $n \geqslant 1$.

Proof by mathematical induction usually consists of the following four steps:

Step 1: **Basis:** Prove the general statement is true for $n = 1$.

Step 2: **Assumption:** Assume the general statement is true for $n = k$.

Step 3: **Inductive**: Show that the general statement is then true for $n = k + 1$.

Step 4: **Conclusion**: The general statement is then true for all positive integers, n.

Example 1

Prove by the method of mathematical induction, that, for $n \in \mathbb{Z}^+$, $\sum_{r=1}^{n}(2r - 1) = n^2$.

$n = 1$; LHS $= \sum_{r=1}^{1}(2r - 1) = 2(1) - 1 = 1$

RHS $= 1^2 = 1$

As LHS = RHS, the summation formula is true for $n = 1$.

1. Basis step:
Substitute $n = 1$ into both the LHS and RHS of the formula to check to see if the formula works for $n = 1$.

Assume that the summation formula is true for $n = k$.

i.e. $\sum_{r=1}^{k}(2r - 1) = k^2$.

2. Assumption step:
In this step you assume that the general statement given is true for $n = k$.

With $n = k + 1$ terms the summation formula becomes:

3. Inductive step:
Sum to k terms plus the $(k + 1)$th term.

$$\sum_{r=1}^{k+1}(2r - 1) = 1 + 3 + \ldots + (2k - 1) + (2(k + 1) - 1)$$

Sum of first k terms is k^2.

This is the $(k + 1)$th term.

$$= k^2 + (2(k + 1) - 1)$$
$$= k^2 + (2k + 2 - 1)$$
$$= k^2 + 2k + 1$$
$$= (k + 1)^2$$

This is the same expression as n^2 with n replaced by $k + 1$.

Therefore, summation formula is true when $n = k + 1$.

If the summation formula is true for $n = k$ then it is shown to be true for $n = k + 1$. As the result is true for $n = 1$, it is now also true for all $n \geq 1$ and $n \in \mathbb{Z}^+$ by mathematical induction.

4. Conclusion step.
Result is true for $n = 1$ and steps two and three imply result is then true for $n = 2$. Continuing to apply steps two and three imply result is true for $n = 3, 4, 5, \ldots$ etc.

Note: $n \in \mathbb{Z}^+$ means that n is a positive integer.

- In step two, it has been **assumed** (but not yet been proved) that the general statement is true for $n = k$, i.e. $\sum_{r=1}^{k}(2r - 1) = k^2$.
- The assumption has then been used in step three *to show* that the general statement is true when $n = k + 1$. At this point we are now able to apply step four, the conclusion.

- In step one, it has been found that the general statement is true for $n = 1$.

- In step two and step three, it has been found that for all $n \in \mathbb{Z}^+$,

  $\sum_{r=1}^{n}(2r - 1)$ is assumed true for $n = k \Rightarrow \sum_{r=1}^{n}(2r - 1)$ is true for $n = k + 1$ **(*)**

- By applying step one and (*), it can be deduced that the general statement is true for $n = 2$. By applying (*) again it can be deduced that the general statement is true for $n = 3$. By continuing to apply (*) it can be deduced that the general statement is true for $n = 3$, $n = 4$, $n = 5$, ... , $n = k$, $n = k + 1$, $n = k + 2$, etc. Therefore the general statement is true for all $n \geqslant 1$ and $n \in \mathbb{Z}^+$ by mathematical induction.

- Therefore the general statement $\sum_{r=1}^{n}(2r - 1) = n^2$ has been proved to be true for any positive integer by mathematical induction.

- Note that proof by induction is *not* used, however, to derive a general statement from first principles. Proof by induction is used to check whether a given statement is true.

Example 2

Prove by the method of mathematical induction, that, for $n \in \mathbb{Z}^+$, $\sum_{r=1}^{n} r^2 = \frac{1}{6}n(n + 1)(2n + 1)$.

$n = 1$; LHS $= \sum_{r=1}^{1} r^2 = 1^2 = 1$

RHS $= \frac{1}{6}(1)(2)(3) = \frac{6}{6} = 1$

As LHS = RHS, the summation formula is true for $n = 1$.

1. Basis step:
Substitute $n = 1$ into both the LHS and RHS of the formula to check to see if the formula works for $n = 1$.

Assume that the summation formula is true for $n = k$.

i.e. $\sum_{r=1}^{k} r^2 = \frac{1}{6}k(k + 1)(2k + 1)$.

2. Assumption step:
In this step you assume that the result given in the question is true for $n = k$.

With $n = k + 1$ terms the summation formula becomes:

3. Inductive step:

$$\sum_{r=1}^{k+1} r^2 = 1^2 + 2^2 + 3^2 + \dots + k^2 + (k+1)^2$$

Sum to k terms plus the $(k + 1)$th term.

$$= \frac{1}{6}k(k+1)(2k+1) + (k+1)^2$$
$$= \frac{1}{6}(k+1)[k(2k+1) + 6(k+1)]$$
$$= \frac{1}{6}(k+1)[2k^2 + k + 6k + 6]$$
$$= \frac{1}{6}(k+1)[2k^2 + 7k + 6]$$
$$= \frac{1}{6}(k+1)(k+2)(2k+3)$$
$$= \frac{1}{6}(k+1)(k+1+1)(2(k+1)+1)$$

This is the same expression as $\frac{1}{6}n(n + 1)(2n + 1)$ with n replaced by $k + 1$.

Therefore, summation formula is true when $n = k + 1$.

If the summation formula is true for $n = k$, then it is shown to be true for $n = k + 1$. As the result is true for $n = 1$, it is now also true for all $n \geq 1$ and $n \in \mathbb{Z}^+$ by mathematical induction.

4. Conclusion step:
Result is true for $n = 1$ and steps two and three imply result is then true for $n = 2$. Continuing to apply steps two and three imply result is true for $n = 3, 4, 5, \dots$ etc.

Example 3

Prove by the method of mathematical induction, that, for $n \in \mathbb{Z}^+$, $\sum_{r=1}^{n} r2^r = 2[1 + (n-1)2^n]$.

$n = 1$; LHS $= \sum_{r=1}^{1} r2^r = 1(2)^1 = 2$

1. Basis step

RHS $= 2[1 + (1-1)2^0] = 2(1) = 2$

As LHS = RHS, the summation formula is true for $n = 1$.

Assume that the summation formula is true for $n = k$.

2. Assumption step

i.e. $\sum_{r=1}^{k} r2^r = 2[1 + (k-1)2^k]$.

With $n = k + 1$ terms the summation formula becomes:

3. Inductive step

$$\sum_{r=1}^{k+1} r2^r = 1(2^1) + 2(2^2) + 3(2^3) + \dots + k(2^k) + (k+1)2^{k+1}$$
$$= 2[1 + (k-1)2^k] + (k+1)2^{k+1}$$
$$= 2 + 2(k-1)2^k + (k+1)2^{k+1}$$

$2^1 \times 2^k = 2^{k+1}$

$$= 2 + (k-1)2^{k+1} + (k+1)2^{k+1}$$
$$= 2 + (k - 1 + k + 1)2^{k+1}$$
$$= 2 + 2k2^{k+1}$$
$$= 2(1 + k2^{k+1})$$
$$= 2[1 + ((k+1) - 1)2^{k+1}]$$

This is the same expression as $2[1 + (n-1)2^n]$ with n replaced by $k + 1$.

Therefore, summation formula is true when $n = k + 1$.

4. Conclusion step

If the summation formula is true for $n = k$, then it is shown to be true for $n = k + 1$. As the result is true for $n = 1$, it is now also true for all $n \geq 1$ and $n \in \mathbb{Z}^+$ by mathematical induction.

Exercise 6A

Prove by the method of mathematical induction, the following statements for $n \in \mathbb{Z}^+$.

1 $\sum_{r=1}^{n} r = \frac{1}{2}n(n+1)$

2 $\sum_{r=1}^{n} r^3 = \frac{1}{4}n^2(n+1)^2$

3 $\sum_{r=1}^{n} r(r-1) = \frac{1}{3}n(n+1)(n-1)$

4 $(1 \times 6) + (2 \times 7) + (3 \times 8) + \ldots + n(n+5) = \frac{1}{3}n(n+1)(n+8)$

5 $\sum_{r=1}^{n} r(3r-1) = n^2(n+1)$

6 $\sum_{r=1}^{n} (2r-1)^2 = \frac{1}{3}n(4n^2-1)$

7 $\sum_{r=1}^{n} 2^r = 2^{n+1} - 2$

8 $\sum_{r=1}^{n} 4^{r-1} = \frac{4^n - 1}{3}$

9 $\sum_{r=1}^{n} r(r!) = (n+1)! - 1$

10 $\sum_{r=1}^{2n} r^2 = \frac{1}{3}n(2n+1)(4n+1)$

6.2 You can use proof by induction to prove that an expression is divisible by a certain integer.

Example 4

Prove, by induction that $3^{2n} + 11$ is divisible by 4 for all positive integers n.

Let $f(n) = 3^{2n} + 11$, where $n \in \mathbb{Z}^+$.

$\therefore f(1) = 3^{2(1)} + 11 = 9 + 11 = 20$, which is divisible by 4.

1. Basis step

$\therefore f(n)$ is divisible by 4 when $n = 1$.

Assume that for $n = k$,

2. Assumption step

$f(k) = 3^{2k} + 11$ is divisible by 4 for $k \in \mathbb{Z}^+$.

$\therefore f(k+1) = 3^{2(k+1)} + 11$

3. Inductive step

$= 3^{2k}.3^2 + 11$

$= 9(3^{2k}) + 11$

$\therefore f(k+1) - f(k) = [9(3^{2k}) + 11] - [3^{2k} + 11]$

$= 8(3^{2k})$

$= 4(2.(3^{2k}))$

$\therefore f(k+1) = f(k) + 4(2.(3^{2k}))$

As both $f(k)$ and $4(2.(3^{2k}))$ are divisible by 4 then the sum of these two must also be divisible by 4.

Therefore $f(n)$ is divisible by 4 when $n = k + 1$.

4. Conclusion step

If $f(n)$ is divisible by 4 when $n = k$, then it has been shown that $f(n)$ is also divisible by 4 when $n = k + 1$. As $f(n)$ is divisible by 4 when $n = 1$, $f(n)$ is also divisible by 4 for all $n \geqslant 1$ and $n \in \mathbb{Z}^+$ by mathematical induction.

Example 5

Prove, by induction that $n^3 - 7n + 9$ is divisible by 3 for all integers for $n \in \mathbb{Z}^+$.

Let $f(n) = n^3 - 7n + 9$, where $n \in \mathbb{Z}^+$.

$f(1) = 1 - 7 + 9 = 3$, which is divisible by 3.

$\therefore$ $f(n)$ is divisible by 3 when $n = 1$.

1. Basis step

Assume that for $n = k$,

$f(k) = k^3 - 7k + 9$ is divisible by 3 for $k \in \mathbb{Z}^+$.

2. Assumption step

$$\begin{aligned} \therefore f(k+1) &= (k+1)^3 - 7(k+1) + 9 \\ &= k^3 + 3k^2 + 3k + 1 - 7(k+1) + 9 \\ &= k^3 + 3k^2 + 3k + 1 - 7k - 7 + 9 \\ &= k^3 + 3k^2 - 4k + 3 \end{aligned}$$

3. Inductive step

Use of Binomial Theorem or multiplying out three brackets

$$\begin{aligned} \therefore f(k+1) - f(k) &= [k^3 + 3k^2 - 4k + 3] - [k^3 - 7k + 9] \\ &= 3k^2 + 3k - 6 \\ &= 3(k^2 + k - 2) \end{aligned}$$

$$\therefore f(k+1) = f(k) + 3(k^2 + k - 2)$$

As both $f(k)$ and $3(k^2 + k - 2)$ are divisible by 3 then the sum of these two must also be divisible by 3.

Therefore $f(n)$ is divisible by 3 when $n = k + 1$.

4. Conclusion step

If $f(n)$ is divisible by 3 when $n = k$, then it has been shown that $f(n)$ is also divisible by 3 when $n = k + 1$. As $f(n)$ is divisible by 3 when $n = 1$, $f(n)$ is also divisible by 3 for all $n \geqslant 1$ and $n \in \mathbb{Z}^+$ by mathematical induction.

Example 6 illustrates a more demanding problem involving divisibility.

Example 6

Prove, by induction that $11^{n+1} + 12^{2n-1}$ is divisible by 133 for all positive integers n.

Let $f(n) = 11^{n+1} + 12^{2n-1}$, where $n \in \mathbb{Z}^+$.

$\therefore f(1) = 11^2 + 12 = 133$, which is divisible by 133.

$\therefore f(n)$ is divisible by 133 when $n = 1$.

1. Basis step

Assume that for $n = k$,

$f(k) = 11^{k+1} + 12^{2k-1}$ is divisible by 133 for $k \in \mathbb{Z}^+$.

2. Assumption step

$$\therefore f(k+1) = 11^{k+1+1} + 12^{2(k+1)-1}$$
$$= 11^{k+1}(11)^1 + 12^{2k-1}(12)^2$$
$$= 11(11^{k+1}) + 144(12^{2k-1})$$

3. Inductive step

$12^{2(k+1)-1} = 12^{2k+2-1}$
$= 12^{2k-1+2}$
$= 12^{2k-1}(12)^2$

$$\therefore f(k+1) - f(k) = [11(11^{k+1}) + 144(12^{2k-1})] - [11^{k+1} + 12^{2k-1}]$$
$$= 10(11^{k+1}) + 143(12^{2k-1})$$
$$= 10(11^{k+1}) + 10(12^{2k-1}) + 133(12^{2k-1})$$
$$= 10[11^{k+1} + 12^{2k-1}] + 133(12^{2k-1})$$

$$\therefore f(k+1) = f(k) + 10[11^{k+1} + 12^{2k-1}] + 133(12^{2k-1})$$
$$= f(k) + 10f(k) + 133(12^{2k-1})$$
$$= 11f(k) + 133(12^{2k-1})$$

As both $11f(k)$ and $133(12^{2k-1})$ are divisible by 133 then the sum of these two must also be divisible by 133.

Therefore $f(n)$ is divisible by 133 when $n = k + 1$.

If $f(n)$ is divisible by 133 when $n = k$, then it has been shown that $f(n)$ is also divisible by 133 when $n = k + 1$. As $f(n)$ is divisible by 133 when $n = 1$, $f(n)$ is also divisible by 133 for all $n \geqslant 1$ and $n \in \mathbb{Z}^+$ by mathematical induction.

4. Conclusion step

Exercise 6B

In Questions 1–8, use the method of mathematical induction to prove the following statements for $n \in \mathbb{Z}^+$.

1 $8^n - 1$ is divisible by 7

2 $3^{2n} - 1$ is divisible by 8

3 $5^n + 9^n + 2$ is divisible by 4

4 $2^{4n} - 1$ is divisible by 15

5 $3^{2n-1} + 1$ is divisible by 4

6 $n^3 + 6n^2 + 8n$ is divisible by 3

7 $n^3 + 5n$ is divisible by 6

8 $2^n.3^{2n} - 1$ is divisible by 17

9 $f(n) = 13^n - 6^n, n \in \mathbb{Z}^+$.

a Express for $k \in \mathbb{Z}^+$, $f(k + 1) - 6f(k)$ in terms of k, simplifying your answer.

b Use the method of mathematical induction to prove that $f(n)$ is divisible by 7 for all $n \in \mathbb{Z}^+$.

10 $g(n) = 5^{2n} - 6n + 8, n \in \mathbb{Z}^+$.

a Express for $k \in \mathbb{Z}^+$, $g(k + 1) - 25g(k)$ in terms of k, simplifying your answer.

b Use the method of mathematical induction to prove that $g(n)$ is divisible by 9 for all $n \in \mathbb{Z}^+$.

11 Use the method of mathematical induction to prove that $8^n - 3^n$ is divisible by 5 for all $n \in \mathbb{Z}^+$.

12 Use the method of mathematical induction to prove that $3^{2n+2} + 8n - 9$ is divisible by 8 for all $n \in \mathbb{Z}^+$.

13 Use the method of mathematical induction to prove that $2^{6n} + 3^{2n-2}$ is divisible by 5 for all $n \in \mathbb{Z}^+$.

6.3 You can use mathematical induction to produce a proof for a general term of a recurrence relation.

In Core 1, you saw recurrence formulae which allowed you to generate successive terms of a sequence.

Example 7

A sequence can be described by the recurrence formula

$$u_{n+1} = 3u_n + 4, n \geqslant 1, u_1 = 1.$$

a Find the first five terms of the sequence.

b Show that the general statement $u_n = 3^n - 2, n \geqslant 1$, gives the same first five terms of the sequence.

a	$u_{n+1} = 3u_n + 4$, $n \geqslant 1$, $u_1 = 1$.
	Substituting $n = 1$; $u_2 = 3u_1 + 4 = 3(1) + 4 = 7$
	Substituting $n = 2$; $u_3 = 3u_2 + 4 = 3(7) + 4 = 25$
	Substituting $n = 3$; $u_4 = 3u_3 + 4 = 3(25) + 4 = 79$
	Substituting $n = 4$; $u_5 = 3u_4 + 4 = 3(79) + 4 = 241$
	The first five terms of the sequence are 1, 7, 25, 79, 241.
b	General statement, $u_n = 3^n - 2$, $n \geqslant 1$
	Substituting $n = 1$; $u_1 = 3^1 - 2 = 3 - 2 = 1$
	Substituting $n = 2$; $u_2 = 3^2 - 2 = 9 - 2 = 7$
	Substituting $n = 3$; $u_3 = 3^3 - 2 = 27 - 2 = 25$
	Substituting $n = 4$; $u_4 = 3^4 - 2 = 81 - 2 = 79$
	Substituting $n = 5$; $u_5 = 3^5 - 2 = 243 - 2 = 241$
	The first five terms of the sequence are ***also*** 1, 7, 25, 79, 241.

Example 8 demonstrates how to apply the method of proof by induction to show that the **general statement** $u_n = 3^n - 2$ is true for the ***recurrence formula*** $u_{n+1} = 3u_n + 4$, with first term $u_1 = 1$ and $n \geqslant 1$.

Example 8

Given that $u_{n+1} = 3u_n + 4$, $u_1 = 1$, prove by induction that $u_n = 3^n - 2$.

$n = 1$; $u_1 = 3^1 - 2 = 1$, as given.	**1. Basis step**
$n = 2$; $u_2 = 3^2 - 2 = 7$, from the general statement.	
and $u_2 = 3u_1 + 4 = 3(1) + 4 = 7$, from the recurrence relation.	
So u_n is true when $n = 1$ and is also true when $n = 2$.	
Assume that for $n = k$ $u_k = 3^k - 2$ is true for $k \in \mathbb{Z}^+$.	**2. Assumption step**
Then $u_{k+1} = 3u_k + 4$	
$= 3(3^k - 2) + 4$	**3. Inductive step** Substituting $u_k = 3^k - 2$
$= 3^{k+1} - 6 + 4$	
$= 3^{k+1} - 2$	This is the same as $u_n = 3^n - 2$ with n replaced by $k + 1$.
Therefore, the general statement, u_n is true when $n = k + 1$.	
If u_n is true when $n = k$, then it has been shown that u_n is also	
true when $n = k + 1$. As u_n is true for $n = 1$ and $n = 2$ then	**4. Conclusion step**
u_n is true for all $n \geqslant 1$ and $n \in \mathbb{Z}^+$ by mathematical induction.	

In the **basis step** of the proof, the general statement was checked for both $n = 1$ and $n = 2$. This is because the first application of the recurrence formula $u_{n+1} = 3u_n + 4$ yields u_2 by using the first given term $u_1 = 1$.

Example 9

Given that $u_{n+2} = 5u_{n+1} - 6u_n$, $u_1 = 13$, $u_2 = 35$, prove by induction that $u_n = 2^{n+1} + 3^{n+1}$.

1. Basis step
The first application of the recurrence formula yields u_3 by using u_1 and u_2. So check general statement for $n = 1, 2, 3$.

$n = 1$; $u_1 = 2^2 + 3^2 = 13$, as given.

$n = 2$; $u_2 = 2^3 + 3^3 = 35$, as given.

$n = 3$; $u_3 = 2^4 + 3^4 = 97$, from the general statement,

and $u_3 = 5u_2 - 6u_1 = 5(35) - 6(13)$

$= 175 - 78 = 97$, from the recurrence relation.

So u_n is true when $n = 1$, $n = 2$ and also when $n = 3$.

2. Assumption step

Assume that, for $n = k$ and $n = k + 1$, that both

$u_k = 2^{k+1} + 3^{k+1}$ and

$u_{k+1} = 2^{k+1+1} + 3^{k+1+1} = 2^{k+2} + 3^{k+2}$ are true for $k \in \mathbb{Z}^+$.

3. Inductive step

Then $u_{k+2} = 5u_{k+1} - 6u_k$

$= 5(2^{k+2} + 3^{k+2}) - 6(2^{k+1} + 3^{k+1})$

$= 5(2^{k+2}) + 5(3^{k+2}) - 6(2^{k+1}) - 6(3^{k+1})$

$= 5(2^{k+2}) + 5(3^{k+2}) - 3(2^{k+2}) - 2(3^{k+2})$

$= 2(2^{k+2}) + 3(3^{k+2})$

$= 2^{k+3} + 3^{k+3}$

$= 2^{k+2+1} + 3^{k+2+1}$

$6(2^{k+1}) = 3(2)(2^{k+1}) = 3(2^{k+2})$

$6(3^{k+1}) = 2(3)(3^{k+1}) = 2(3^{k+2})$

This is the same as $u_n = 2^{n+1} + 3^{n+1}$ with n replaced by $k + 1$.

Therefore, the general statement, u_n is true when $n = k + 2$.

4. Conclusion step

If u_n is true when $n = k$ and $n = k + 1$ then it has been shown that u_n is also true when $n = k + 2$. As u_n is true for $n = 1$, $n = 2$ and $n = 3$, then u_n is true for all $n \geqslant 1$ and $n \in \mathbb{Z}^+$ by mathematical induction.

In Example 9, each term of the sequence depends upon the two previous terms. Therefore u_n was assumed to be true for both $n = k$ and $n = k + 1$.

Exercise 6C

1 Given that $u_{n+1} = 5u_n + 4$, $u_1 = 4$, prove by induction that $u_n = 5^n - 1$.

2 Given that $u_{n+1} = 2u_n + 5$, $u_1 = 3$, prove by induction that $u_n = 2^{n+2} - 5$.

3 Given that $u_{n+1} = 5u_n - 8$, $u_1 = 3$, prove by induction that $u_n = 5^{n-1} + 2$.

4 Given that $u_{n+1} = 3u_n + 1$, $u_1 = 1$, prove by induction that $u_n = \frac{3^n - 1}{2}$.

5 Given that $u_{n+2} = 5u_{n+1} - 6u_n$, $u_1 = 1$, $u_2 = 5$ prove by induction that $u_n = 3^n - 2^n$.

6 Given that $u_{n+2} = 6u_{n+1} - 9u_n$, $u_1 = -1$, $u_2 = 0$, prove by induction that $u_n = (n - 2)3^{n-1}$.

7 Given that $u_{n+2} = 7u_{n+1} - 10u_n$, $u_1 = 1$, $u_2 = 8$, prove by induction that $u_n = 2(5^{n-1}) - 2^{n-1}$.

8 Given that $u_{n+2} = 6u_{n+1} - 9u_n$, $u_1 = 3$, $u_2 = 36$, prove by induction that $u_n = (3n - 2)3^n$.

6.4 You can use proof by induction to prove general statements involving matrix multiplication.

Example 10

Use mathematical induction to prove that $\begin{pmatrix} 1 & -1 \\ 0 & 2 \end{pmatrix}^n = \begin{pmatrix} 1 & 1 - 2^n \\ 0 & 2^n \end{pmatrix}$ for $n \in \mathbb{Z}^+$.

1. Basis step

$n = 1$; LHS $= \begin{pmatrix} 1 & -1 \\ 0 & 2 \end{pmatrix}^1 = \begin{pmatrix} 1 & -1 \\ 0 & 2 \end{pmatrix}$

RHS $= \begin{pmatrix} 1 & 1 - 2^1 \\ 0 & 2^1 \end{pmatrix} = \begin{pmatrix} 1 & -1 \\ 0 & 2 \end{pmatrix}$

Substitute $n = 1$ into both the LHS and RHS of the formula to check to see if the formula works for $n = 1$.

As LHS = RHS, the matrix equation is true for $n = 1$.

2. Assumption step

Assume that the matrix equation is true for $n = k$.

i.e. $\begin{pmatrix} 1 & -1 \\ 0 & 2 \end{pmatrix}^k = \begin{pmatrix} 1 & 1 - 2^k \\ 0 & 2^k \end{pmatrix}$

In this step you assume that the general statement given is true for $n = k$.

3. Inductive step

With $n = k + 1$ the matrix equation becomes

$$\begin{pmatrix} 1 & -1 \\ 0 & 2 \end{pmatrix}^{k+1} = \begin{pmatrix} 1 & -1 \\ 0 & 2 \end{pmatrix}^k \begin{pmatrix} 1 & -1 \\ 0 & 2 \end{pmatrix}$$

$$= \begin{pmatrix} 1 & 1 - 2^k \\ 0 & 2^k \end{pmatrix} \begin{pmatrix} 1 & -1 \\ 0 & 2 \end{pmatrix}$$

Using the assumption step.

$$= \begin{pmatrix} 1 + 0 & -1 + 2 - 2(2^k) \\ 0 + 0 & 0 + 2(2^k) \end{pmatrix}$$

$$= \begin{pmatrix} 1 & 1 - 2^{k+1} \\ 0 & 2^{k+1} \end{pmatrix}$$

The matrix equation is now true when n is replaced by $k + 1$.

Therefore the matrix equation is true when $n = k + 1$.

4. Conclusion step

If the matrix equation is true for $n = k$, then it is shown to be true for $n = k + 1$. As the matrix equation is true for $n = 1$, it is now also true for all $n \geqslant 1$ and $n \in \mathbb{Z}^+$ by mathematical induction.

Example 11

Use mathematical induction to prove that $\begin{pmatrix} -2 & 9 \\ -1 & 4 \end{pmatrix}^n = \begin{pmatrix} -3n+1 & 9n \\ -n & 3n+1 \end{pmatrix}$ for $n \in \mathbb{Z}^+$.

$n = 1$; LHS $= \begin{pmatrix} -2 & 9 \\ -1 & 4 \end{pmatrix}^1 = \begin{pmatrix} -2 & 9 \\ -1 & 4 \end{pmatrix}$

RHS $= \begin{pmatrix} -3(1)+1 & 9(1) \\ -(1) & 3(1)+1 \end{pmatrix} = \begin{pmatrix} -2 & 9 \\ -1 & 4 \end{pmatrix}$

As LHS = RHS, the matrix equation is true for $n = 1$.

1. Basis step

Substitute $n = 1$ into both the LHS and RHS of the formula to check to see if the formula works for $n = 1$.

Assume that the matrix equation is true for $n = k$.

i.e. $\begin{pmatrix} -2 & 9 \\ -1 & 4 \end{pmatrix}^k = \begin{pmatrix} -3k+1 & 9k \\ -k & 3k+1 \end{pmatrix}$.

2. Assumption step

In this step you assume that the general statement given is true for $n = k$.

With $n = k + 1$ the matrix equation becomes

$$\begin{pmatrix} -2 & 9 \\ -1 & 4 \end{pmatrix}^{k+1} = \begin{pmatrix} -2 & 9 \\ -1 & 4 \end{pmatrix}^k \begin{pmatrix} -2 & 9 \\ -1 & 4 \end{pmatrix}$$

$$= \begin{pmatrix} -3k+1 & 9k \\ -k & 3k+1 \end{pmatrix} \begin{pmatrix} -2 & 9 \\ -1 & 4 \end{pmatrix}$$

$$= \begin{pmatrix} -6k-2-9k & -27k+9+36k \\ 2k-3k-1 & -9k+12k+4 \end{pmatrix}$$

$$= \begin{pmatrix} -3k-2 & 9k+9 \\ -k-1 & 3k+4 \end{pmatrix}$$

$$= \begin{pmatrix} -3(k+1)+1 & 9(k+1) \\ -(k+1) & 3(k+1)+1 \end{pmatrix}$$

3. Inductive step

Using the assumption step.

The matrix equation is now true when n is replaced by $k + 1$.

Therefore the matrix equation is true when $n = k + 1$.

If the matrix equation is true for $n = k$, then it is shown to be true for $n = k + 1$. As the matrix equation is true for $n = 1$, it is now also true for all $n \geqslant 1$ and $n \in \mathbb{Z}^+$ by mathematical induction.

4. Conclusion step

Exercise 6D

Prove by the method of mathematical induction the following statements for $n \in \mathbb{Z}^+$.

1 $\begin{pmatrix} 1 & 2 \\ 0 & 1 \end{pmatrix}^n = \begin{pmatrix} 1 & 2n \\ 0 & 1 \end{pmatrix}$

2 $\begin{pmatrix} 3 & -4 \\ 1 & -1 \end{pmatrix}^n = \begin{pmatrix} 2n+1 & -4n \\ n & -2n+1 \end{pmatrix}$

3 $\begin{pmatrix} 2 & 0 \\ 1 & 1 \end{pmatrix}^n = \begin{pmatrix} 2^n & 0 \\ 2^n - 1 & 1 \end{pmatrix}$

4 $\begin{pmatrix} 5 & -8 \\ 2 & -3 \end{pmatrix}^n = \begin{pmatrix} 4n+1 & -8n \\ 2n & 1-4n \end{pmatrix}$

5 $\begin{pmatrix} 2 & 5 \\ 0 & 1 \end{pmatrix}^n = \begin{pmatrix} 2^n & 5(2^n - 1) \\ 0 & 1 \end{pmatrix}$

Mixed exercise 6E

1 Prove by induction that $9^n - 1$ is divisible by 8 for $n \in \mathbb{Z}^+$.

2 The matrix $\mathbf{B}$ is given by $\mathbf{B} = \begin{pmatrix} 1 & 0 \\ 0 & 3 \end{pmatrix}$.

a Find $\mathbf{B}^2$ and $\mathbf{B}^3$.

b Hence write down a general statement for $\mathbf{B}^n$, for $n \in \mathbb{Z}^+$.

c Prove, by induction that your answer to part **b** is correct.

3 Prove by induction that for $n \in \mathbb{Z}^+$, that $\sum_{r=1}^{n}(3r + 4) = \frac{1}{2}n(3n + 11)$.

4 A sequence $u_1, u_2, u_3, u_4, \ldots$ is defined by $u_{n+1} = 5u_n - 3(2^n)$, $u_1 = 7$.

a Find the first four terms of the sequence.

b Prove, by induction for $n \in \mathbb{Z}^+$, that $u_n = 5^n + 2^n$.

5 The matrix $\mathbf{A}$ is given by $\mathbf{A} = \begin{pmatrix} 9 & 16 \\ -4 & -7 \end{pmatrix}$.

a Prove by induction that $\mathbf{A}^n = \begin{pmatrix} 8n + 1 & 16n \\ -4n & 1 - 8n \end{pmatrix}$ for $n \in \mathbb{Z}^+$.

The matrix $\mathbf{B}$ is given by $\mathbf{B} = (\mathbf{A}^n)^{-1}$

b Hence find $\mathbf{B}$ in terms of n.

6 The function f is defined by $f(n) = 5^{2n-1} + 1$, where $n \in \mathbb{Z}^+$.

a Show that $f(n + 1) - f(n) = \mu(5^{2n-1})$, where μ is an integer to be determined.

b Hence prove by induction that $f(n)$ is divisible by 6.

7 Use the method of mathematical induction to prove that $7^n + 4^n + 1$ is divisible by 6 for all $n \in \mathbb{Z}^+$.

8 A sequence $u_1, u_2, u_3, u_4, \ldots$ is defined by $u_{n+1} = \frac{3u_n - 1}{4}$, $u_1 = 2$.

a Find the first five terms of the sequence.

b Prove, by induction for $n \in \mathbb{Z}^+$, that $u_n = 4\left(\frac{3}{4}\right)^n - 1$.

9 A sequence $u_1, u_2, u_3, u_4, \ldots$ is defined by $u_n = 3^{2n} + 7^{2n-1}$.

a Show that $u_{n+1} - 9u_n = \lambda(7^{2k-1})$, where λ is an integer to be determined.

b Hence prove by induction that u_n is divisible by 8 for all positive integers n.

10 Prove by induction, for all positive integers n, that

$$(1 \times 5) + (2 \times 6) + (3 \times 7) + \ldots + n(n + 4) = \frac{1}{6}n(n + 1)(2n + 13).$$

Summary of key points

1 Mathematical induction is used to prove whether or not general statements are true, usually for positive integers, n.

2 When performing a proof by mathematical induction you need to apply the following four steps:

- **basis:** Show the general statement if true for $n = 1$.
- **assumption:** Assume that the general statement is true for $n = k$.
- **induction:** Show the general statement is true for $n = k + 1$.
- **conclusion:** Then state that the general statement is then true for all positive integers, n.

3 Proof by induction is of no use for deriving formulae from first principles. Proof by induction is used, however, to check whether or not a general statement is true.

Review Exercise

1 $\mathbf{A} = \begin{pmatrix} 3 & 2 & 1 \\ 0 & 2 & -1 \end{pmatrix}$, $\mathbf{B} = \begin{pmatrix} 2 & 0 \\ 3 & -1 \end{pmatrix}$, $\mathbf{C} = \begin{pmatrix} 4 \\ -3 \\ 1 \end{pmatrix}$

Determine whether or not the following products exist. Where the product exists, evaluate the product. Where the product does not exist, give a reason for this.

a **AB** **b** **BA** **c** **BAC** **d** **CBA**.

2 $\mathbf{M} = \begin{pmatrix} 0 & 3 \\ -1 & 2 \end{pmatrix}$, $\mathbf{I} = \begin{pmatrix} 1 & 0 \\ 0 & 1 \end{pmatrix}$ and $\mathbf{O} = \begin{pmatrix} 0 & 0 \\ 0 & 0 \end{pmatrix}$.

Find the values of the constants a and b such that $\mathbf{M}^2 + a\mathbf{M} + b\mathbf{I} = \mathbf{O}$. (E)

3 $\mathbf{A} = \begin{pmatrix} 4 & 1 \\ 3 & 6 \end{pmatrix}$

Show that $\mathbf{A}^2 - 10\mathbf{A} + 21\mathbf{I} = \mathbf{O}$.

4 $\mathbf{A} = \begin{pmatrix} a & b \\ c & d \end{pmatrix}$.

Find an expression for λ, in terms of a, b, c and d so that $\mathbf{A}^2 - (a + d)\mathbf{A} = \lambda\mathbf{I}$, where $\mathbf{I}$ is the 2×2 unit matrix. (E)

5 $\mathbf{A} = \begin{pmatrix} 2 & 3 \\ p & -1 \end{pmatrix}$, where p is a real constant.

Given that **A** is singular,

a find the value of p.

Given instead that det $(\mathbf{A}) = 4$,

b find the value of p.

Using the value of p found in **b**,

c show that $\mathbf{A}^2 - \mathbf{A} = k\mathbf{I}$, stating the value of the constant k.

6 $\mathbf{A} = \begin{pmatrix} 2 & -1 \\ -3 & 1 \end{pmatrix}$

a Find $\mathbf{A}^{-1}$.

Given that $\mathbf{A}^5 = \begin{pmatrix} 251 & -109 \\ -327 & 142 \end{pmatrix}$,

b find $\mathbf{A}^4$.

7 A triangle T, of area 18 cm^2, is transformed into a triangle T' by the matrix **A** where,

$\mathbf{A} = \begin{pmatrix} k & k-1 \\ -3 & 2k \end{pmatrix}$, $k \in \mathbb{R}$.

a Find det (**A**), in terms of k.

Given that the area of T' is 198 cm^2,

b find the possible values of k.

8 A linear transformation from $\mathbb{R}^2 \to \mathbb{R}^2$ is defined by $\mathbf{p} = \mathbf{Nq}$, where $\mathbf{N}$ is a 2×2 matrix and $\mathbf{p}$, $\mathbf{q}$ are 2×1 column vectors.

Given that $\mathbf{p} = \begin{pmatrix} 3 \\ 7 \end{pmatrix}$ when $\mathbf{q} = \begin{pmatrix} 1 \\ 0 \end{pmatrix}$, and that $\mathbf{p} = \begin{pmatrix} 6 \\ -1 \end{pmatrix}$ when $\mathbf{q} = \begin{pmatrix} 2 \\ -3 \end{pmatrix}$, find $\mathbf{N}$.

E

9 $\mathbf{A} = \begin{pmatrix} 4 & -1 \\ -6 & 2 \end{pmatrix}$, $\mathbf{B}^{-1} = \begin{pmatrix} 2 & 0 \\ 3 & p \end{pmatrix}$

a Find $\mathbf{A}^{-1}$.

b Find $(\mathbf{AB})^{-1}$, in terms of p.

Given also that $\mathbf{AB} = \begin{pmatrix} -1 & 2 \\ 3 & -4 \end{pmatrix}$,

c find the value of p.

10 $\mathbf{A} = \begin{pmatrix} 2 & -1 \\ 7 & -3 \end{pmatrix}$

a Show that $\mathbf{A}^3 = \mathbf{I}$.

b Deduce that $\mathbf{A}^2 = \mathbf{A}^{-1}$.

c Use matrices to solve the simultaneous equations
$2x - y = 3$,
$7x - 3y = 2$.

11 $\mathbf{A} = \begin{pmatrix} 5 & -2 \\ 5 & 5 \end{pmatrix}$, $\mathbf{B} = \begin{pmatrix} 4 & 2 \\ 5 & 1 \end{pmatrix}$

a Find $\mathbf{A}^{-1}$.

b Show that $\mathbf{A}^{-1}\mathbf{BA} = \begin{pmatrix} \lambda_1 & 0 \\ 0 & \lambda_2 \end{pmatrix}$, stating the values of the constants λ_1 and λ_2.

12 $\mathbf{A} = \begin{pmatrix} 4p & -q \\ -3p & q \end{pmatrix}$, where p and q are non-zero constants.

a Find $\mathbf{A}^{-1}$, in terms of p and q.

Given that $\mathbf{AX} = \begin{pmatrix} 2p & 3q \\ -p & q \end{pmatrix}$,

b find $\mathbf{X}$, in terms of p and q.

13 $\mathbf{A} = \begin{pmatrix} 4 & 2 \\ 5 & 3 \end{pmatrix}$, $\mathbf{B} = \begin{pmatrix} 3 & -1 \\ -4 & 5 \end{pmatrix}$

Find

a $\mathbf{AB}$,

b $\mathbf{AB} - \mathbf{BA}$.

Given that $\mathbf{C} = \mathbf{AB} - \mathbf{BA}$,

c find $\mathbf{C}^2$,

d give a geometrical interpretation of the transformation represented by $\mathbf{C}^2$.

14 The matrix $\mathbf{A}$ represents reflection in the x-axis.

The matrix $\mathbf{B}$ represents a rotation of 135°, in the anti-clockwise direction, about (0, 0).

Given that $\mathbf{C} = \mathbf{AB}$,

a find the matrix $\mathbf{C}$,

b show that $\mathbf{C}^2 = \mathbf{I}$.

15 The linear transformation $T: \mathbb{R}^2 \to \mathbb{R}^2$ is represented by the matrix $\mathbf{M}$, where

$\mathbf{M} = \begin{pmatrix} a & b \\ c & d \end{pmatrix}$.

The transformation T maps the point with coordinates (1, 0) to the point with coordinates (3, 2) and the point with coordinates (2, 1) to the point with coordinates (2, 1).

a Find the values of a, b, c and d.

b Show that $\mathbf{M}^2 = \mathbf{I}$.

The transformation T maps the point with coordinates (p, q) to the point with coordinates $(8, -3)$.

c Find the value of p and the value of q.

16 The linear transformation T is defined by

$\begin{pmatrix} x \\ y \end{pmatrix} \to \begin{pmatrix} 2y - x \\ 3y \end{pmatrix}$.

The linear transformation T is represented by the matrix $\mathbf{C}$.

a Find $\mathbf{C}$.

The quadrilateral $OABC$ is mapped by T to the quadrilateral $OA'B'C'$, where the coordinates of A', B' and C' are (0, 3), (10, 15) and (10, 12) respectively.

b Find the coordinates of A, B and C.

c Sketch the quadrilateral $OABC$ and verify that $OABC$ is a rectangle.

17 $\mathbf{A} = \begin{pmatrix} 3 & -2 \\ -1 & 4 \end{pmatrix}$, $\mathbf{B} = \begin{pmatrix} 0.8 & -0.4 \\ 0.2 & -0.6 \end{pmatrix}$ and $\mathbf{C} = \mathbf{AB}$.

a Find $\mathbf{C}$.

b Give a geometrical interpretation of the transformation represented by $\mathbf{C}$.

The square $OXYZ$, where the coordinates of X and Y are (0, 3) and (3, 3), is transformed into the quadrilateral $OX'Y'Z'$, by the transformation represented by $\mathbf{C}$.

c Find the coordinates of Z'.

18 Given that $\mathbf{A} = \begin{pmatrix} 5 & 3 \\ -2 & -1 \end{pmatrix}$ and $\mathbf{B} = \begin{pmatrix} 1 & 1 \\ 0 & 2 \end{pmatrix}$, find the matrices $\mathbf{C}$ and $\mathbf{D}$ such that

a $\mathbf{AC} = \mathbf{B}$,

b $\mathbf{DA} = \mathbf{B}$.

A linear transformation from $\mathbb{R}^2 \to \mathbb{R}^2$ is defined by the matrix $\mathbf{B}$.

c Prove that the line with equation $y = mx$ is mapped onto another line through the origin O under this transformation.

d Find the gradient of this second line in terms of m. E

19 Referred to an origin O and coordinate axes Ox and Oy, transformations from $\mathbb{R}^2 \to \mathbb{R}^2$ are represented by the matrices $\mathbf{L}$, $\mathbf{M}$ and $\mathbf{N}$, where

$$\mathbf{L} = \begin{pmatrix} 0 & -1 \\ 1 & 0 \end{pmatrix}, \mathbf{M} = \begin{pmatrix} 2 & 0 \\ 0 & 2 \end{pmatrix} \text{ and } \mathbf{N} = \begin{pmatrix} 1 & -1 \\ 1 & 1 \end{pmatrix}.$$

a Explain the geometrical effect of the transformations $\mathbf{L}$ and $\mathbf{M}$.

b Show that $\mathbf{LM} = \mathbf{N}^2$.

The transformation represented by the matrix $\mathbf{N}$ consists of a rotation of angle θ about O, followed by an enlargement, centre O, with positive scale factor k.

c Find the value of θ and the value of k.

d Find $\mathbf{N}^8$. E

20 $\mathbf{A}$, $\mathbf{B}$ and $\mathbf{C}$ are 2×2 matrices.

a Given that $\mathbf{AB} = \mathbf{AC}$, and that $\mathbf{A}$ is not singular, prove that $\mathbf{B} = \mathbf{C}$.

b Given that $\mathbf{AB} = \mathbf{AC}$, where $\mathbf{A} = \begin{pmatrix} 3 & 6 \\ 1 & 2 \end{pmatrix}$ and $\mathbf{B} = \begin{pmatrix} 1 & 5 \\ 0 & 1 \end{pmatrix}$, find a matrix $\mathbf{C}$ whose elements are all non-zero. E

21 Use standard formulae to show that

$$\sum_{r=1}^{n} 3r(r-1) = n(n^2-1).$$

22 Use standard formulae to show that

$$\sum_{r=1}^{n} (r^2-1) = \frac{n}{6}(2n+5)(n-1).$$ E

23 Use standard formulae to show that

$$\sum_{r=1}^{n} (2r-1)^2 = \frac{1}{3}n(4n^2-1).$$ E

24 Use standard formulae to show that

$$\sum_{r=1}^{n} r(r^2-3) = \frac{1}{4}n(n+1)(n-2)(n+3).$$

25 **a** Use standard formulae to show that

$$\sum_{r=1}^{n} r(2r-1) = \frac{n(n+1)(4n-1)}{6}.$$

b Hence, evaluate $\sum_{r=11}^{30} r(2r-1)$. E

26 Evaluate $\sum_{r=0}^{12}(r^2 + 2^r)$. **E**

27 Evaluate $\sum_{r=1}^{50}(r + 1)(r + 2)$. **E**

28 Use standard formulae to show that

$$\sum_{r=1}^{n} r(r^2 - n) = \frac{n^2(n^2 - 1)}{4}.$$

29 **a** Use standard formulae to show that

$$\sum_{r=1}^{n} r(3r + 1) = n(n + 1)^2.$$

b Hence evaluate $\sum_{r=40}^{100} r(3r + 1)$. **E**

30 **a** Show that

$$\sum_{r=1}^{n}(2r - 1)(2r + 3) = \frac{n}{3}(4n^2 + 12n - 1).$$

b Hence find $\sum_{r=5}^{35}(2r - 1)(2r + 3)$. **E**

31 **a** Use standard formulae to show that

$$\sum_{r=1}^{n}(6r^2 + 4r - 5) = n(2n^2 + 5n - 2).$$

b Hence calculate the value of

$$\sum_{r=10}^{25}(6r^2 + 4r - 5).$$ **E**

32 **a** Use standard formulae to show that

$$\sum_{r=1}^{n}(r + 1)(r + 5) = \tfrac{1}{6}n(n + 7)(2n + 7).$$

b Hence calculate the value of

$$\sum_{r=10}^{40}(r + 1)(r + 5).$$ **E**

33 **a** Use standard formulae to show that

$$\sum_{r=1}^{n} r^2(r + 1) = \frac{n(n + 1)(3n^2 + 7n + 2)}{12}.$$

b Find $\sum_{r=4}^{30}(2r)^2(2r + 2)$. **E**

34 Using the formula $\sum_{r=1}^{n} r^2 = \frac{n}{6}(n + 1)(2n + 1)$,

a show that $\sum_{r=1}^{n}(4r^2 - 1) = \frac{n}{3}(4n^2 + 6n - 1)$.

Given that $\sum_{r=1}^{12}(4r^2 + kr - 1) = 2120$,

where k is a constant,

b find the value of k.

35 **a** Use standard formulae to show that

$$\sum_{r=1}^{n} r(3r - 5) = n(n + 1)(n - 2).$$

b Hence show that $\sum_{r=n}^{2n} r(3r - 5) = 7n(n^2 - 1)$.

36 **a** Use standard formulae to show that

$$\sum_{r=1}^{n} r(r + 1) = \tfrac{1}{3}n(n + 1)(n + 2).$$

b Hence, or otherwise, show that

$\sum_{r=n}^{3n} r(r + 1) = \tfrac{1}{3}n(2n + 1)(pn + q)$, stating the values of the integers p and q. **E**

37 Given that $\sum_{r=1}^{n} r^2(r - 1)$

$= \tfrac{1}{12}n(n + 1)(pn^2 + qn + r)$,

a find the values of p, q and r.

b Hence evaluate $\sum_{r=50}^{100} r^2(r - 1)$.

38 **a** Use standard formula to show that

$$\sum_{r=1}^{n} r(r + 2) = \tfrac{1}{6}n(n + 1)(2n + 7).$$

b Hence, or otherwise, find the value of

$$\sum_{r=1}^{10}(r + 2)\log_4 2^r.$$

39 Use the method of mathematical induction to prove that, for all positive integers n, $\sum_{r=1}^{n} \frac{1}{r(r + 1)} = \frac{n}{n + 1}$. **E**

40 Use the method of mathematical induction to prove that

$$\sum_{r=1}^{n} r(r+3) = \tfrac{1}{3}n(n+1)(n+5).$$ **E**

41 Prove by induction that, for $n \in \mathbb{Z}^+$,

$$\sum_{r=1}^{n} (2r-1)^2 = \tfrac{1}{3}n\,(2n-1)(2n+1).$$ **E**

42 The rth term a_r in a series is given by $a_r = r(r+1)(2r+1)$.

Prove, by mathematial induction, that the sum of the first n terms of the series is $\tfrac{1}{2}n(n+1)^2(n+2)$. **E**

43 Prove, by induction, that

$$\sum_{r=1}^{n} r^2(r-1) = \tfrac{1}{12}n(n-1)(n+1)(3n+2).$$ **E**

44 Given that $u_1 = 8$ and $u_{n+1} = 4u_n - 9n$, use mathematical induction to prove that $u_n = 4^n + 3n + 1$, $n \in \mathbb{Z}^+$.

45 Given that $u_1 = 0$ and $u_{r+1} = 2r - u_r$, use mathematical induction to prove that $2u_n = 2n - 1 + (-1)^n$, $n \in \mathbb{Z}^+$. **E**

46 $f(n) = (2n+1)7^n - 1$.

Prove by induction that, for all positive integers n, $f(n)$ is divisible by 4. **E**

47 $\mathbf{A} = \begin{pmatrix} 1 & c \\ 0 & 2 \end{pmatrix}$, where c is a constant.

Prove by induction that, for all positive integers n,

$$\mathbf{A}^n = \begin{pmatrix} 1 & (2^n-1)c \\ 0 & 2^n \end{pmatrix}$$

48 Given that $u_1 = 4$ and that $2u_{r+1} + u_r = 6$, use mathematical induction to prove that

$$u_n = 2 - \left(-\frac{1}{2}\right)^{n-2}, \text{ for } n \in \mathbb{Z}^+.$$

49 Prove by induction that, for all $n \in \mathbb{Z}^+$,

$$\sum_{r=1}^{n} r\left(\tfrac{1}{2}\right)^r = 2 - \left(\tfrac{1}{2}\right)^n(n+2).$$ **E**

50 $\mathbf{A} = \begin{pmatrix} 3 & 1 \\ -4 & -1 \end{pmatrix}$

Prove by induction that, for all positive integers n,

$$\mathbf{A}^n = \begin{pmatrix} 2n+1 & n \\ -4n & -2n+1 \end{pmatrix}$$

51 Given that $f(n) = 3^{4n} + 2^{4n+2}$,

a show that, for $k \in \mathbb{Z}^+$, $f(k+1) - f(k)$ is divisible by 15,

b prove that, for $n \in \mathbb{Z}^+$, $f(n)$ is divisible by 5. **E**

52 $f(n) = 24 \times 2^{4n} + 3^{4n}$, where n is a non-negative integer.

a Write down $f(n+1) - f(n)$.

b Prove, by induction, that $f(n)$ is divisible by 5. **E**

53 Prove that the expression $7^n + 4^n + 1$ is divisible by 6 for all positive integers n. **E**

54 Prove by induction that $4^n + 6n - 1$ is divisible by 9 for $n \in \mathbb{Z}^+$. **E**

55 Prove that the expression $3^{4n-1} + 2^{4n-1} + 5$ is divisible by 10 for all positive integers n. **E**

56 **a** Express $\dfrac{6x+10}{x+3}$ in the form $p + \dfrac{q}{x+3}$, where p and q are integers to be found.

The sequence of real numbers $u_1, u_2, u_3, \ldots$ is such that $u_1 = 5.2$ and $u_{n+1} = \dfrac{6u_n + 10}{u_n + 3}$.

b Prove by induction that $u_n > 5$, for $n \in \mathbb{Z}^+$. **E**

57 Given that $n \in \mathbb{Z}^+$, prove, by mathematical induction, that $2(4^{2n+1}) + 3^{3n+1}$ is divisible by 11. **E**

Examination style paper

1* Use the standard results for $\sum_{r=1}^{n} r$ and for $\sum_{r=1}^{n} r^2$ to show that, for all positive integers n,

$\sum_{r=1}^{n} (r+1)(3r+2) = n(an^2 + bn + c)$, where the values of a, b and c should be stated. (5)

2* $f(x) = x^3 + 3x - 6$

The equation $f(x) = 0$ has a root α in the interval [1, 1.5].

a Taking 1.25 as a first approximation to α, apply the Newton–Raphson procedure once to $f(x)$ to obtain a second approximation to α. Give your answer to three significant figures. (5)

b Show that the answer which you obtained is an accurate estimate to three significant figures. (2)

3*

$$\mathbf{R} = \begin{pmatrix} -\frac{1}{\sqrt{2}} & -\frac{1}{\sqrt{2}} \\ \frac{1}{\sqrt{2}} & -\frac{1}{\sqrt{2}} \end{pmatrix} \text{ and } \mathbf{S} = \begin{pmatrix} \sqrt{2} & 0 \\ 0 & \sqrt{2} \end{pmatrix}$$

a Describe fully the geometric transformation represented by each of **R** and **S.** (4)

b Calculate **RS.** (2)

The unit square, U, is transformed by the transformation represented by **S** followed by the transformation represented by **R**.

c Find the area of the image of U after both transformations have taken place. (2)

4* $f(z) = z^4 + 3z^2 - 6z + 10$

Given that $1 + i$ is a complex root of $f(z) = 0$,

a state a second complex root of this equation. (1)

b Use these two roots to find a quadratic factor of $f(z)$, with real coefficients. (3)

Another quadratic factor of $f(z)$ is $z^2 + 2z + 5$.

c Find the remaining two roots of $f(z) = 0$. (3)

5* The rectangular hyperbola H has equation $xy = c^2$. The points $P\left(cp, \frac{c}{p}\right)$ and $Q\left(cq, \frac{c}{q}\right)$ lie on the hyperbola H.

a Show that the gradient of the chord PQ is $-\frac{1}{pq}$. (3)

The point R, $\left(3c, \frac{c}{3}\right)$ also lies on H and PR is perpendicular to QR.

b Show that this implies that the gradient of the chord PQ is 9. (4)

6*

$$\mathbf{M} = \begin{pmatrix} x & 2x - 7 \\ -1 & x + 4 \end{pmatrix}$$

a Find the inverse of matrix **M**, in terms of x, given that **M** is non-singular. (4)

b Show that **M** is a singular matrix for two values of x and state these values. (4)

7* The complex numbers z and w are given by $z = \frac{7 - i}{1 - i}$ and $w = iz$.

a Express z and w in the form $a + ib$, where a and b are real numbers. (3)

b Find the modulus of w. (1)

c Find the argument of w in radians to two decimal places. (2)

d Show z and w on an Argand diagram (2)

e Find $|z - w|$. (2)

8* The parabola C has equation $y^2 = 16x$.

a Find the equation of the normal to C at the point P, (1, 4). (6)

The normal at P meets the directrix to the parabola at the point Q.

b Find the coordinates of Q. (3)

c Give the coordinates of the point R on the parabola, which is equidistant from Q and from the focus of C. (2)

9* **a** Use the method of mathematical induction to prove that, for $n \in \mathbb{Z}^+$,

$$\sum_{r=1}^{n} r + \left(\tfrac{1}{2}\right)^{r-1} = \tfrac{1}{2}(n^2 + n + 4) - \left(\tfrac{1}{2}\right)^{n-1}.$$ (6)

b $f(n) = 3^{n+2} + (-1)^n 2^n$, $n \in \mathbb{Z}^+$.

By considering $2f(n+1) - f(n)$ and using the method of mathematical induction prove that, for $n \in \mathbb{Z}^+$, $3^{n+2} + (-1)^n 2^n$ is divisible by 5. (6)

Answers

Exercise 1A

1 $13 + 11i$
2 $5 + 2i$
3 $4 + i$
4 $13 + i$
5 -3
6 $9 + 9i$
7 $1 - 4i$
8 $7 - 4i$
9 $4 + 2i$
10 $4i$

11 $12 + 15i$
12 $11 - 5i$
13 0
14 $14 + 4i$
15 $24 - 12i$
16 $7 - 21i$
17 $12 + 5i$
18 $24 + 7i$
19 $3 + 2i$
20 $2\sqrt{2} + 2i$

21 $3i$
22 $7i$
23 $11i$
24 $100i$
25 $15i$
26 $i\sqrt{5}$
27 $2i\sqrt{3}$
28 $3i\sqrt{5}$
29 $10i\sqrt{2}$
30 $7i\sqrt{3}$

31 $x = -1 \pm 2i$
32 $x = 1 \pm 3i$
33 $x = -2 \pm 5i$
34 $x = -5 \pm i$
35 $x = 3 \pm 3i$
36 $x = -2 \pm i\sqrt{3}$
37 $x = 3 \pm i\sqrt{2}$
38 $x = 1 \pm 2i\sqrt{6}$
39 $x = -\frac{5}{2} \pm \frac{5i\sqrt{3}}{2}$
40 $x = \frac{-3 \pm i\sqrt{11}}{2}$

Exercise 1B

1 $11 + 23i$
2 $36 + 33i$
3 $15 + 23i$
4 $2 - 110i$
5 $-5 - 25i$
6 $39 + 80i$
7 $-77 - 36i$
8 $10i$
9 $54 - 62i$
10 $-46 + 9i$
11 -1
12 81
13 $2i$
14 $-60i$
15 16

Exercise 1C

1 **a** $8 - 2i$ **b** $6 + 5i$
c $\frac{2}{3} + \frac{1}{2}i$ **d** $\sqrt{5} - i\sqrt{10}$

2 **a** $z + z^* = 12, zz^* = 45$ **b** $z + z^* = 20, zz^* = 125$
c $z + z^* = \frac{3}{2}, zz^* = \frac{5}{8}$ **d** $z + z^* = 2\sqrt{5}, zz^* = 50$

3 $9 + 8i$
4 $\frac{22}{25} - \frac{21}{25}i$
5 $\frac{37}{10} + \frac{1}{10}i$
6 $\frac{3}{5} + \frac{1}{5}i$
7 $-\frac{6}{5} - \frac{7}{5}i$
8 $-\frac{11}{50} + \frac{27}{50}i$
9 $\frac{31}{2} + \frac{25}{2}i$
10 $\frac{6}{17} - \frac{7}{17}i$
11 $-\frac{31}{2} - \frac{17}{2}i$
12 $\frac{3}{5} + \frac{4}{5}i$
13 $\frac{7}{2} + \frac{1}{2}i$
14 $\frac{41}{5} - \frac{3}{5}i$
15 $\frac{8}{5} + \frac{9}{5}i$
16 $6 + 8i$

17 **a** $-1 + 5i, -1 - 5i$ **b** -2 **c** 26

18 **a** $4 + 3i, 4 - 3i$ **b** 8 **c** 25

19 $x^2 - 4x + 13 = 0$

20 $x^2 + 10x + 41 = 0$

Exercise 1D

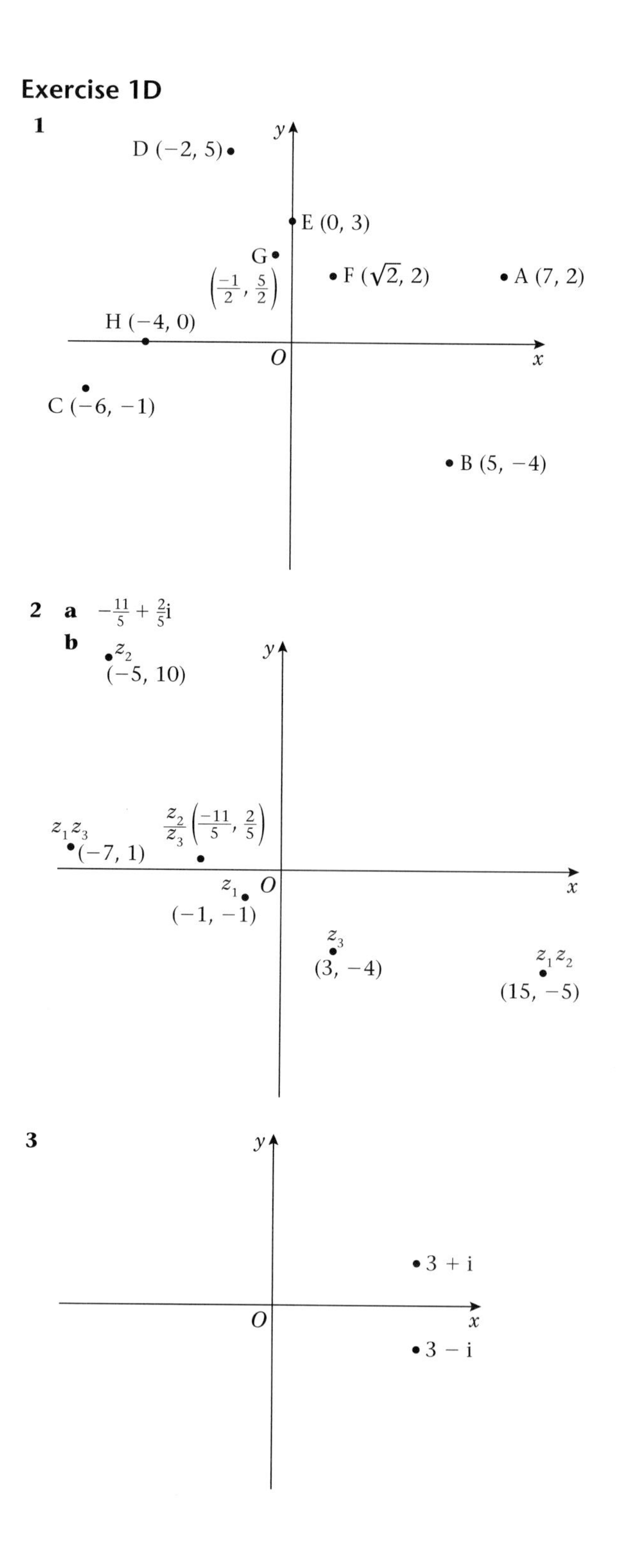

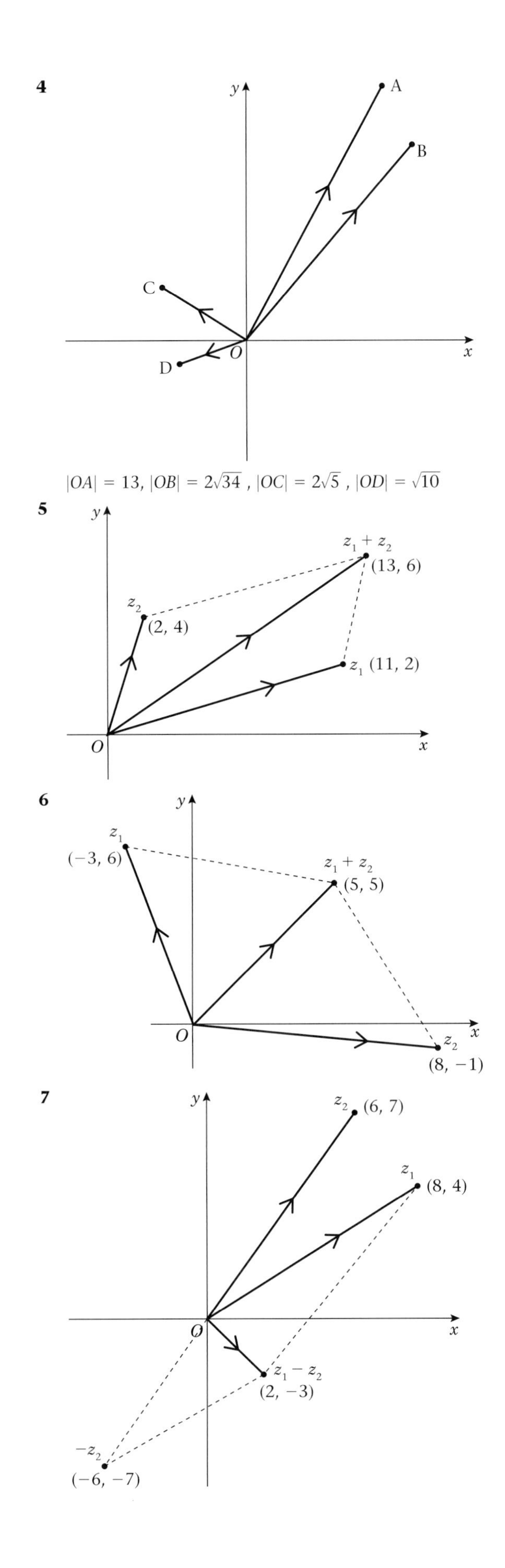

$|OA| = 13$, $|OB| = 2\sqrt{34}$, $|OC| = 2\sqrt{5}$, $|OD| = \sqrt{10}$

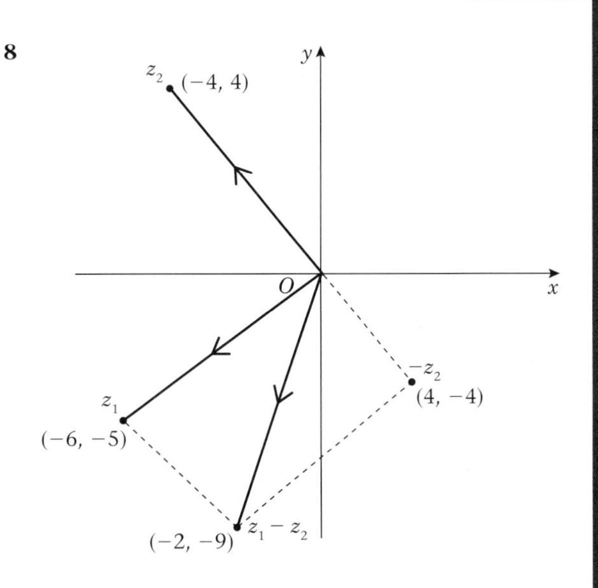

Exercise 1E

1 Modulus = 13, argument = 0.39

2 Modulus = 2, argument = $\frac{\pi}{6}$

3 Modulus = $3\sqrt{5}$, argument = 2.03

4 Modulus = $2\sqrt{2}$, argument = $-\frac{\pi}{4}$

5 Modulus = $\sqrt{113}$, argument = −2.42

6 Modulus = $\sqrt{137}$, argument = 1.92

7 Modulus = $\sqrt{15}$, argument = −0.46

8 Modulus = 17, argument = −2.06

Exercise 1F

1 **a** $2\sqrt{2}\left(\cos\frac{\pi}{4} + \text{i}\sin\frac{\pi}{4}\right)$

b $3\left(\cos\frac{\pi}{2} + \text{i}\sin\frac{\pi}{2}\right)$

c $5(\cos 2.21 + \text{i}\sin 2.21)$

d $2\left(\cos -\frac{\pi}{3} + \text{i}\sin -\frac{\pi}{3}\right)$

e $\sqrt{29}\,(\cos(-1.95) + \text{i}\sin(-1.95))$

f $20(\cos\pi + \text{i}\sin\pi)$

g $25(\cos(-1.29) + \text{i}\sin(-1.29))$

h $5\sqrt{2}\left(\cos\frac{3\pi}{4} + \text{i}\sin\frac{3\pi}{4}\right)$

2 **a** $\frac{3}{2}\left(\cos\left(-\frac{\pi}{3}\right) + \text{i}\sin\left(-\frac{\pi}{3}\right)\right)$

b $\frac{\sqrt{5}}{5}(\cos 0.46 + \text{i}\sin 0.46)$

c $1\left(\cos\frac{\pi}{2} + \text{i}\sin\frac{\pi}{2}\right)$

3 **a** $3 + 3\text{i}$

b $-3\sqrt{2} + 3\sqrt{2}\text{i}$

c $\frac{\sqrt{3}}{2} + \frac{3}{2}\text{i}$

d -7i

e $-2\sqrt{3} - 2\text{i}$

4 **a** $|z_1| = 5$, $|z_2| = 5$, $z_1 z_2 = 24 + 7\text{i}$

b $|z_1| = \sqrt{5}$, $|z_2| = 2\sqrt{5}$, $z_1 z_2 = -8 + 6\text{i}$

c $|z_1| = 13$, $|z_2| = 25$, $z_1 z_2 = -253 + 204\text{i}$

d $|z_1| = \sqrt{5}$, $|z_2| = \sqrt{5}$, $z_1 z_2 = -2\sqrt{6} + \text{i}$

Exercise 1G

1 $a = 3, b = \frac{1}{2}$
2 $a = 7, b = -2$
3 $a = -3, b = 7$
4 3
5 $a = 2$ and $b = 6$, or $a = -2$ and $b = -6$
6 $x = \frac{3}{13}, y = \frac{2}{13}$
7 $x = \frac{3}{2}, y = -\frac{1}{2}$
8 $x = -1, y = 1$, modulus $= \sqrt{2}$, argument $= \dfrac{3\pi}{4}$
9 $\pm(4 + 3\text{i})$
10 $\pm(6 + 5\text{i})$
11 $\pm(3 - 2\text{i})$
12 $\pm(1 + \text{i})$

Exercise 1H

1 $x^2 - 2x + 5 = 0$
2 $x^2 - 6x + 34 = 0$
3 $x^2 - 2ax + a^2 + 16 = 0$
4 Roots are -1, $-4 + 3\text{i}$ and $-4 - 3\text{i}$
5 Roots are 3, $-\frac{1}{2} + \frac{1}{2}\text{i}$ and $-\frac{1}{2} - \frac{1}{2}\text{i}$
6 Roots are $-\frac{1}{2}$, $-\frac{1}{2} + \dfrac{\sqrt{3}}{2}\text{i}$ and $-\frac{1}{2} - \dfrac{\sqrt{3}}{2}\text{i}$
7 Roots are 4, $-4 + \text{i}$ and $-4 - \text{i}$
8 Roots are 3, -3, $6 + 2\text{i}$ and $6 - 2\text{i}$
9 Roots are $2 + 3\text{i}$, $2 - 3\text{i}$, $-3 + \text{i}$ and $-3 - \text{i}$
10

y
2i
−2
O
2
x
−2i

11 $a = 1, b = 0, c = 2, d = 4, e = -8, f = 16$

Mixed exercise 1I

1 **a** $-1 - 4\text{i}$, $-1 + 4\text{i}$
b

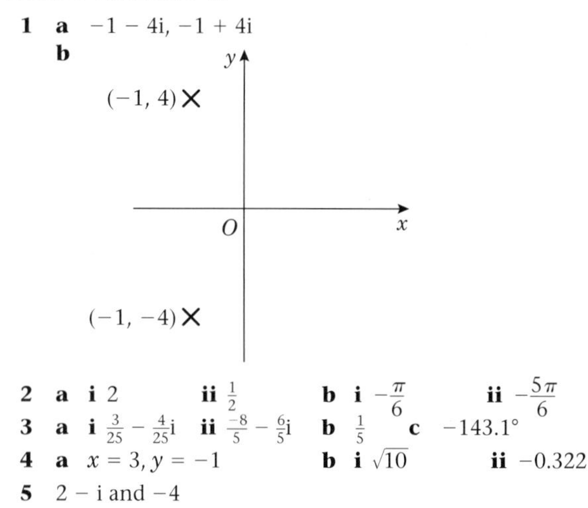

2 **a** **i** 2 **ii** $\frac{1}{2}$ **b** **i** $-\dfrac{\pi}{6}$ **ii** $-\dfrac{5\pi}{6}$
3 **a** **i** $\frac{3}{25} - \frac{4}{25}\text{i}$ **ii** $\frac{-8}{5} - \frac{6}{5}\text{i}$ **b** $\frac{1}{5}$ **c** $-143.1°$
4 **a** $x = 3, y = -1$ **b** **i** $\sqrt{10}$ **ii** -0.322
5 $2 - \text{i}$ and -4

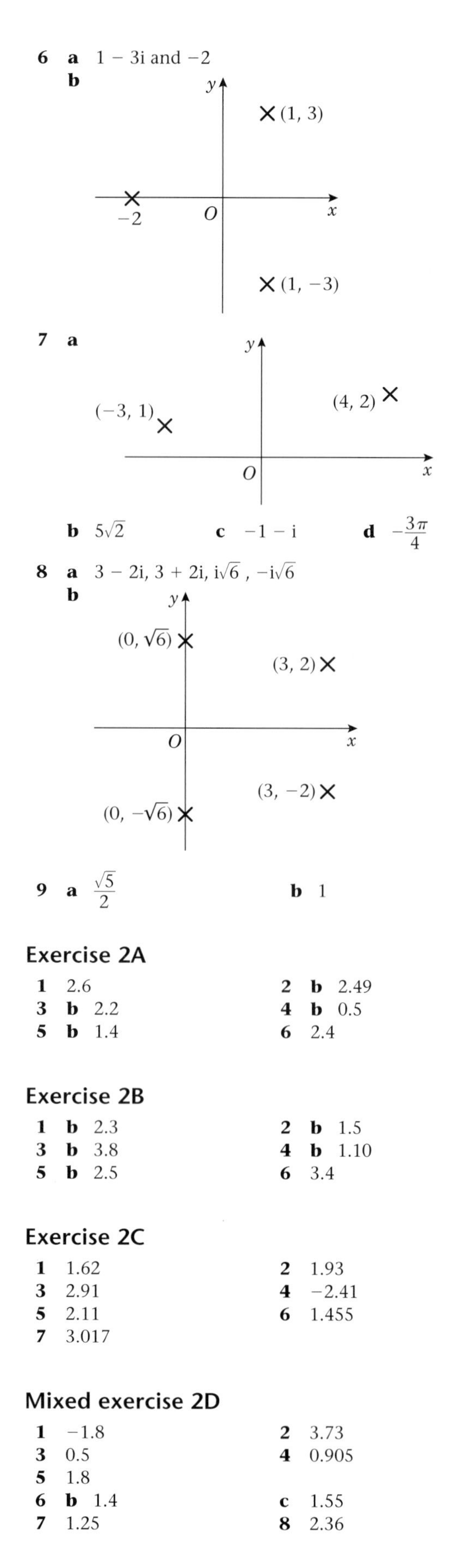

6 **a** $1 - 3\text{i}$ and -2
b

7 **a**

b $5\sqrt{2}$ **c** $-1 - \text{i}$ **d** $-\dfrac{3\pi}{4}$

8 **a** $3 - 2\text{i}$, $3 + 2\text{i}$, $\text{i}\sqrt{6}$, $-\text{i}\sqrt{6}$
b

9 **a** $\dfrac{\sqrt{5}}{2}$ **b** 1

Exercise 2A

1 2.6 **2** **b** 2.49
3 **b** 2.2 **4** **b** 0.5
5 **b** 1.4 **6** 2.4

Exercise 2B

1 **b** 2.3 **2** **b** 1.5
3 **b** 3.8 **4** **b** 1.10
5 **b** 2.5 **6** 3.4

Exercise 2C

1 1.62 **2** 1.93
3 2.91 **4** -2.41
5 2.11 **6** 1.455
7 3.017

Mixed exercise 2D

1 -1.8 **2** 3.73
3 0.5 **4** 0.905
5 1.8
6 **b** 1.4 **c** 1.55
7 1.25 **8** 2.36

Exercise 3A

1

t	-4	-3	-2	-1	-0.5	0	0.5	1	2	3	4
$x = 2t^2$	32	18	8	2	0.5	0	0.5	2	8	18	32
$y = 4t$	-16	-12	-8	-4	-2	0	2	4	8	12	16

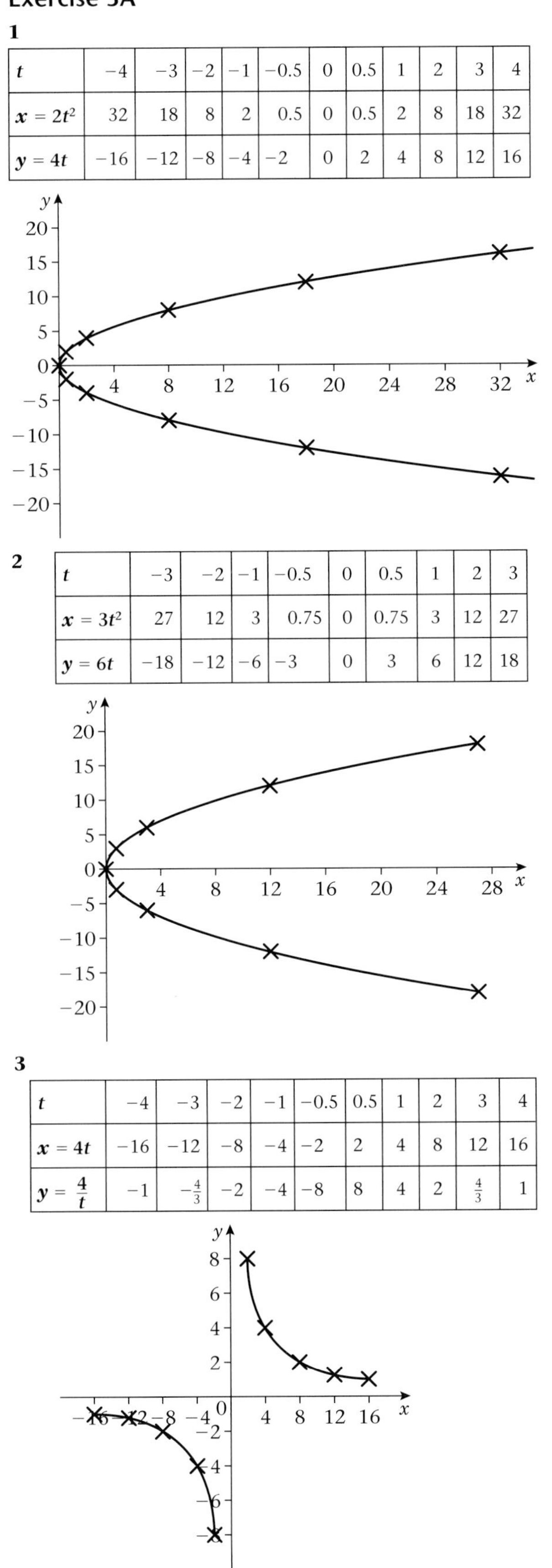

2

t	-3	-2	-1	-0.5	0	0.5	1	2	3
$x = 3t^2$	27	12	3	0.75	0	0.75	3	12	27
$y = 6t$	-18	-12	-6	-3	0	3	6	12	18

3

t	-4	-3	-2	-1	-0.5	0.5	1	2	3	4
$x = 4t$	-16	-12	-8	-4	-2	2	4	8	12	16
$y = \frac{4}{t}$	-1	$-\frac{4}{3}$	-2	-4	-8	8	4	2	$\frac{4}{3}$	1

4 **a** $y^2 = 20x$ **b** $y^2 = 2x$
c $y^2 = 200x$ **d** $y^2 = \frac{4}{5}x$
e $y^2 = 10x$ **f** $y^2 = 4\sqrt{3}x$
g $x^2 = 8y$ **h** $x^2 = 12y$

5 **a** $xy = 1$ **b** $xy = 49$
c $xy = 45$ **d** $xy = \frac{1}{25}$

6 **a** $xy = 9$
b

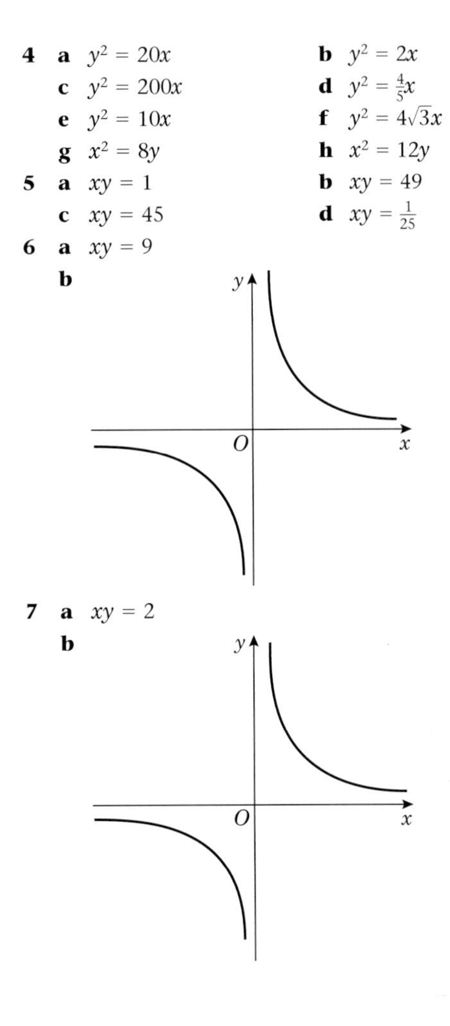

7 **a** $xy = 2$
b

Exercise 3B

1 **a** $y^2 = 20x$ **b** $y^2 = 32x$
c $y^2 = 4x$ **d** $y^2 = 6x$
e $y^2 = 2\sqrt{3}x$

2 **a** $(3, 0); x + 3 = 0$ **b** $(5, 0); x + 5 = 0$
c $\left(\frac{5}{2}, 0\right); x + \frac{5}{2} = 0$ **d** $(\sqrt{3}, 0); x + \sqrt{3} = 0$
e $\left(\frac{\sqrt{2}}{4}, 0\right); x + \frac{\sqrt{2}}{4} = 0$ **f** $\left(\frac{5\sqrt{2}}{4}, 0\right); x + \frac{5\sqrt{2}}{4} = 0$

5 **b** $(0, 2); y + 2 = 0$
c

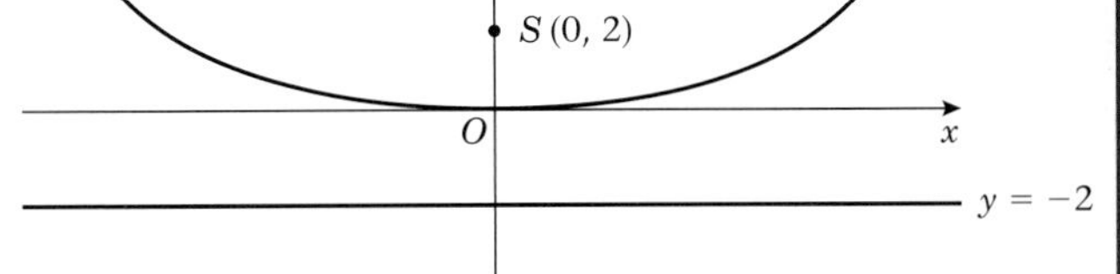

Exercise 3C

1 $(3, 3)$ and $\left(\frac{3}{4}, -\frac{3}{2}\right)$

2 $16\sqrt{2}$

3 $(25, 5)$

4 **a** $y^2 = 24x$ **b** $(6, 0)$; $x + 6 = 0$

c

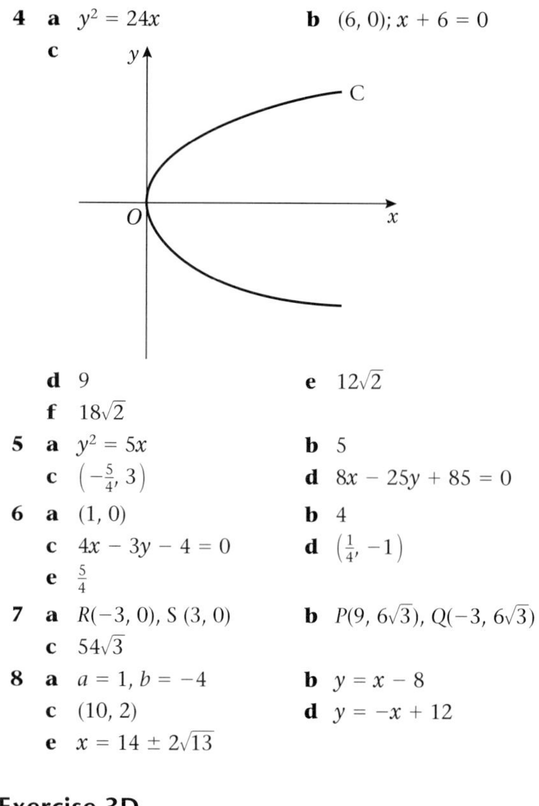

d 9 **e** $12\sqrt{2}$

f $18\sqrt{2}$

5 **a** $y^2 = 5x$ **b** 5

c $\left(-\frac{5}{4}, 3\right)$ **d** $8x - 25y + 85 = 0$

6 **a** $(1, 0)$ **b** 4

c $4x - 3y - 4 = 0$ **d** $\left(\frac{1}{4}, -1\right)$

e $\frac{5}{4}$

7 **a** $R(-3, 0)$, $S(3, 0)$ **b** $P(9, 6\sqrt{3})$, $Q(-3, 6\sqrt{3})$

c $54\sqrt{3}$

8 **a** $a = 1, b = -4$ **b** $y = x - 8$

c $(10, 2)$ **d** $y = -x + 12$

e $x = 14 \pm 2\sqrt{13}$

Exercise 3D

1 **a** $x - 4y + 16 = 0$ **b** $\sqrt{2}x - 2y + 4\sqrt{2} = 0$

c $x + y - 10 = 0$ **d** $16x + y - 16 = 0$

e $x + 2y + 7 = 0$ **f** $2x + y - 8\sqrt{2} = 0$

2 **a** $x + y - 15 = 0$ **b** $2x - 8y - 45 = 0$

3 **a** 4 **b** $y = -x + 12$

c $(36, -24)$ **d** $32\sqrt{2}$

4 **a** $x - 8y - 126 = 0$ **b** $\left(128, \frac{1}{4}\right)$

5 **b** $(6\sqrt{2}, 4\sqrt{2})$ or $(-6\sqrt{2}, -4\sqrt{2})$

6 **a** $xy = 3$ **b** $8x - 2y - 15\sqrt{3} = 0$

c $\left(-\frac{1}{8}\sqrt{3}, -8\sqrt{3}\right)$

7 **a** $\frac{1}{2}$; $(1, 4)$ **b** $(-15, 0)$

c $(-1, 0)$ **d** 28

Exercise 3E

3 **a** 5 **c** $\frac{25}{2}t^3$

4 **b** $(a, -2a)$ and $(16a, 8a)$

5 **b** $(-4, 5)$ **c** $(8, 2)$ and $\left(-\frac{8}{5}, -10\right)$

d $x + 4y - 16 = 0$; $25x + 4y + 80 = 0$

6 **a** $(-at^2, 0)$ **b** $(2a + at^2, 0)$

c $2a^2t(1 + t^2)$

7 **b** $(0, 0)$, $(8, 8)$ and $(8, -8)$

c $y = 0$, $2x + y - 24 = 0$ and $2x - y - 24 = 0$

8 **a** $(0, at)$ **b** $(a, 0)$

9 **b** -6 **c** $(24, 24)$ and $\left(\frac{3}{2}, -6\right)$

Mixed exercise 3F

1 **a** $(3, 0)$ **b** $\left(\frac{4}{3}, 4\right)$ **c** 6

2 **a** $\frac{3}{2}$ **b** $(6, 0)$ **d** 30

3 **a** $y^2 = 48x$ **b** $x + 12 = 0$

c $(16, 16\sqrt{3})$ **d** $96\sqrt{3}$

4 **a** $(1, 4)$ and $(64, 32)$

c $x + 2y - 9 = 0$ and $4x + y - 288 = 0$

d $(81, -36)$

e $9\sqrt{97}$

5 **a** focus of $C(a, 0)$, $Q(-a, 0)$

b $(a, 2a)$ or $(a, -2a)$

6 **b** $4x - y = 45$

c $\left(-\frac{3}{4}, -48\right)$

7 $x + 4y - 12 = 0$ and $x + 4y + 12 = 0$

8 **a** $X(2ct, 0)$ and $Y\left(0, \frac{2c}{t}\right)$

b $6\sqrt{2}$

9 **b** $4ty = x + 16at^2$ **c** $(8at^2, 6at)$

10 **c** $\frac{c^2}{2a}$ **d** $y = \frac{c^2x}{4a^2}$

e $\frac{8a}{5}$ **g** $\frac{4a}{5}$

Review exercise 1

1 **a** $|z_1 z_2^*| = 5\sqrt{5}$, $\tan\arg(z_1 z_2^*) = -\frac{1}{2}$

b $\left|\frac{z_1}{z_2}\right| = \frac{\sqrt{5}}{5}$, $\tan\arg\left(\frac{z_1}{z_2}\right) = -\frac{1}{2}$

2 **a** $\frac{1}{2}$ **b** $-\frac{1}{4}$

3 **b** $\frac{3\pi}{4}$

4 **a** $\frac{1}{3} - \frac{4}{3}\mathrm{i}$

b

y
$\times z^*\left(\frac{1}{3}, \frac{4}{3}\right)$
O
x
$\times z\left(\frac{1}{3}, -\frac{4}{3}\right)$

c $z = \frac{\sqrt{17}}{3}\cos(-76°) + \mathrm{i}\frac{\sqrt{17}}{3}\sin(-76°)$

$z^* = \frac{\sqrt{17}}{3}\cos 76° + \mathrm{i}\frac{\sqrt{17}}{3}\sin 76°$

5 **a** **i** $\frac{2\pi}{3}$ **ii** $\frac{\pi}{6}$ **b** $0 + \mathrm{i}$; $\frac{\pi}{2}$

6 **a** $2 - \mathrm{i}$ and $-2 + \mathrm{i}$

b

y
$(-2, 1)\times$
O
x
$\times (2, -1)$

7 **a**

y
$(-9, 17) \times$
O
x

b 2.06 **c** $1 - 2\mathrm{i}$

8 **a** $-1 + 2\mathrm{i}$ **b** 2.03

9 **b** $\frac{\sqrt{2}}{2}$; $-\frac{3\pi}{4}$

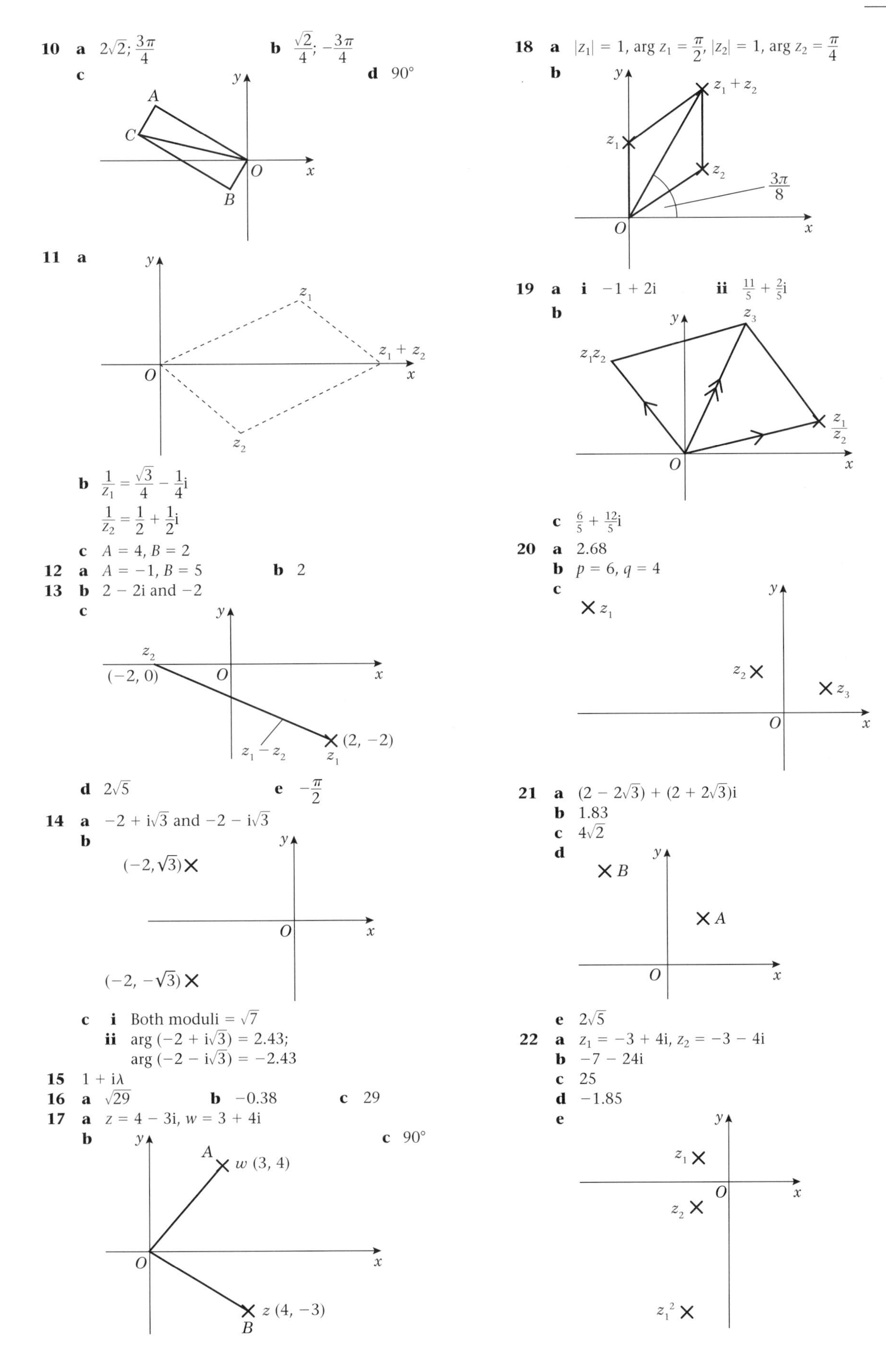

10 **a** $2\sqrt{2}; \frac{3\pi}{4}$ **b** $\frac{\sqrt{2}}{4}; -\frac{3\pi}{4}$

c

d 90°

11 **a**

b $\frac{1}{z_1} = \frac{\sqrt{3}}{4} - \frac{1}{4}\mathrm{i}$

$\frac{1}{z_2} = \frac{1}{2} + \frac{1}{2}\mathrm{i}$

c $A = 4, B = 2$

12 **a** $A = -1, B = 5$ **b** 2

13 **b** $2 - 2\mathrm{i}$ and -2

c

d $2\sqrt{5}$ **e** $-\frac{\pi}{2}$

14 **a** $-2 + \mathrm{i}\sqrt{3}$ and $-2 - \mathrm{i}\sqrt{3}$

b

c **i** Both moduli $= \sqrt{7}$

ii $\arg(-2 + \mathrm{i}\sqrt{3}) = 2.43$;
$\arg(-2 - \mathrm{i}\sqrt{3}) = -2.43$

15 $1 + \mathrm{i}\lambda$

16 **a** $\sqrt{29}$ **b** -0.38 **c** 29

17 **a** $z = 4 - 3\mathrm{i}, w = 3 + 4\mathrm{i}$

b

c 90°

18 **a** $|z_1| = 1, \arg z_1 = \frac{\pi}{2}, |z_2| = 1, \arg z_2 = \frac{\pi}{4}$

b

19 **a** **i** $-1 + 2\mathrm{i}$ **ii** $\frac{11}{5} + \frac{2}{5}\mathrm{i}$

b

c $\frac{6}{5} + \frac{12}{5}\mathrm{i}$

20 **a** 2.68

b $p = 6, q = 4$

c

21 **a** $(2 - 2\sqrt{3}) + (2 + 2\sqrt{3})\mathrm{i}$

b 1.83

c $4\sqrt{2}$

d

e $2\sqrt{5}$

22 **a** $z_1 = -3 + 4\mathrm{i}, z_2 = -3 - 4\mathrm{i}$

b $-7 - 24\mathrm{i}$

c 25

d -1.85

e

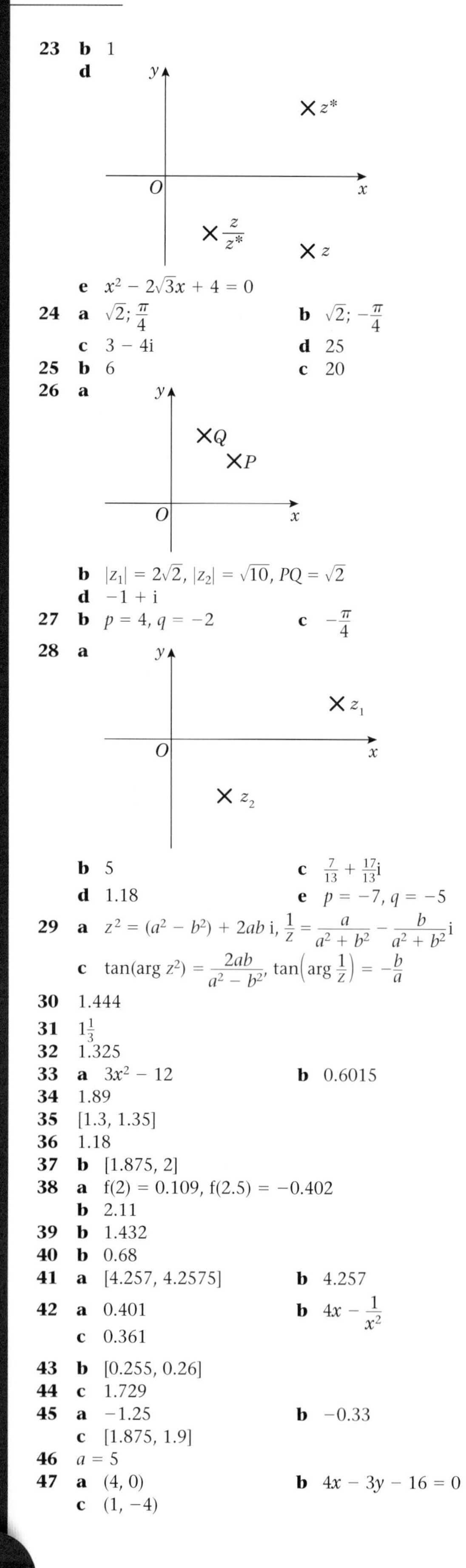

23 **b** 1
d

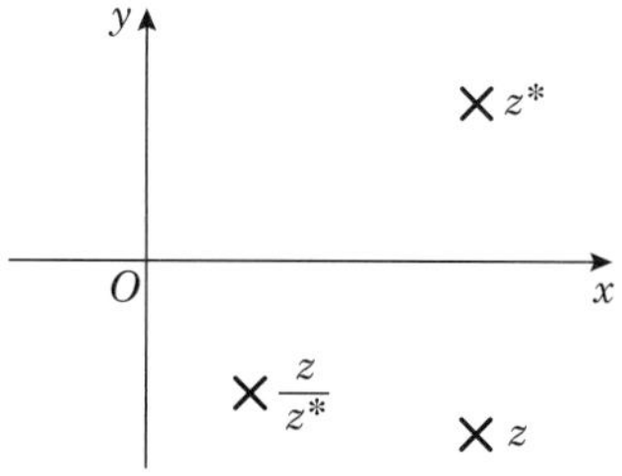

e $x^2 - 2\sqrt{3}x + 4 = 0$

24 **a** $\sqrt{2}; \frac{\pi}{4}$ **b** $\sqrt{2}; -\frac{\pi}{4}$
c $3 - 4\mathrm{i}$ **d** 25

25 **b** 6 **c** 20

26 **a**

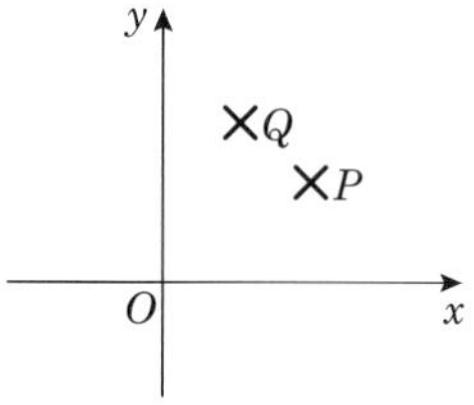

b $|z_1| = 2\sqrt{2}, |z_2| = \sqrt{10}, PQ = \sqrt{2}$
d $-1 + \mathrm{i}$

27 **b** $p = 4, q = -2$ **c** $-\frac{\pi}{4}$

28 **a**

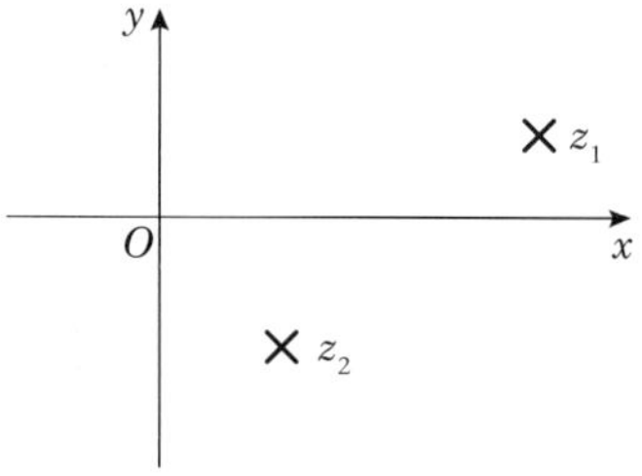

b 5 **c** $\frac{7}{13} + \frac{17}{13}\mathrm{i}$
d 1.18 **e** $p = -7, q = -5$

29 **a** $z^2 = (a^2 - b^2) + 2ab\,\mathrm{i}, \frac{1}{z} = \frac{a}{a^2 + b^2} - \frac{b}{a^2 + b^2}\mathrm{i}$
c $\tan(\arg z^2) = \frac{2ab}{a^2 - b^2}, \tan\left(\arg \frac{1}{z}\right) = -\frac{b}{a}$

30 1.444

31 $1\frac{1}{3}$

32 1.325

33 **a** $3x^2 - 12$ **b** 0.6015

34 1.89

35 [1.3, 1.35]

36 1.18

37 **b** [1.875, 2]

38 **a** f(2) = 0.109, f(2.5) = −0.402
b 2.11

39 **b** 1.432

40 **b** 0.68

41 **a** [4.257, 4.2575] **b** 4.257

42 **a** 0.401 **b** $4x - \frac{1}{x^2}$
c 0.361

43 **b** [0.255, 0.26]

44 **c** 1.729

45 **a** −1.25 **b** −0.33
c [1.875, 1.9]

46 $a = 5$

47 **a** (4, 0) **b** $4x - 3y - 16 = 0$
c (1, −4)

48 **a**

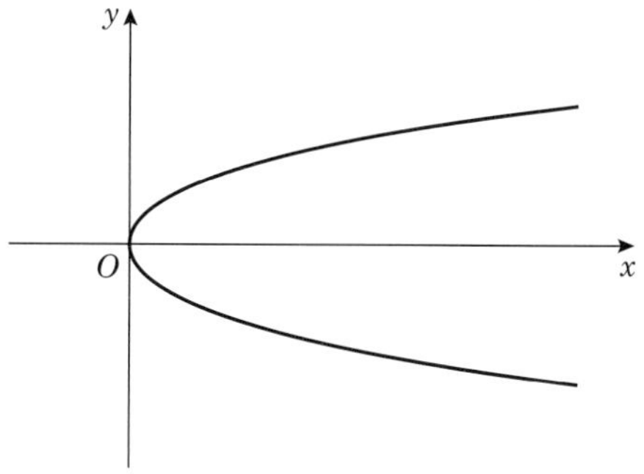

b $60\sqrt{2}$

49 $4\sqrt{15}$

50 **a** 8 **b** $y = 2x + 4$
c 4

51 **a** $y = \frac{4}{5}x + \frac{8}{5}$ **b** (−5, −2.4)

52 **a** $x + 4y = 24$ **b** $6\sqrt{7}$

53 **a** $5x - 4y - 9 = 0$ **b** (−3.2, −6.25)

54 (−8, 2)

55 **a** $t = \frac{1}{2}, P(6, 24)$ **b** $y = 2x + 12$
c $y = -4x + 48$

56 **b** $(3\sqrt{6}, 4\sqrt{6})$ and $(-3\sqrt{6}, -4\sqrt{6})$

59 **b** $\left(-\frac{4}{3}, -12\right)$ and $\left(12, \frac{4}{3}\right)$

61 **b** $(apq, a(p + q))$ **c** $p = 4 - q$

62 **b** $\left(a\left(\frac{t^2 + 2}{t}\right)^2, -2a\left(\frac{t^2 + 2}{t}\right)\right)$

63 **b** $\left(-\frac{c}{t^3}, -ct^3\right)$

64 **c** 1

65 **b** $(2cp, 0)$ **c** $\frac{c}{p}(1 + p^4)^{\frac{1}{2}}$
d $\left(\frac{c}{3}, 3c\right)$

66 **b** $q = -p - \frac{2}{p}$ **c** $\frac{2}{3}$

67 **a** (8, 0) **b** $x = -8$

Exercise 4A

1 **a** 2 × 2 **b** 2 × 1 **c** 2 × 3
d 1 × 3 **e** 1 × 2 **f** 3 × 3

2 **a** $\begin{pmatrix} 8 & -1 \\ 1 & 4 \end{pmatrix}$ **b** $\begin{pmatrix} 2 & 2 \\ -2 & -5 \end{pmatrix}$
c $\begin{pmatrix} 0 & 0 \\ 0 & 0 \end{pmatrix}$

3 **a** Not possible **b** $\begin{pmatrix} -2 \\ 3 \end{pmatrix}$
c $(1 \ \ 1 \ \ 4)$ **d** Not possible
e $(3 \ \ -1 \ \ 4)$ **f** Not possible
g $(-3 \ \ 1 \ \ -4)$

4 $a = 6, b = 3, c = 2, d = -1$

5 $a = 4, b = 3, c = 5$

6 $a = 2, b = -2, c = 2, d = 1, e = -1, f = 3$

Exercise 4B

1 **a** $\begin{pmatrix} 6 & 0 \\ 12 & -18 \end{pmatrix}$ **b** $\begin{pmatrix} 1 & 0 \\ 2 & -3 \end{pmatrix}$ **c** $\begin{pmatrix} 2 \\ -2 \end{pmatrix}$

2 $k = 3, x = -1$

3 $a = 3, b = -3.5, c = -1, d = 2$

4 $a = 5, b = 5, c = -2, d = 2$

5 $k = \frac{3}{2}$

Exercise 4C

1 **a** 1×2 **b** 3×3 **c** 1×2
d 2×2 **e** 2×3 **f** 3×2

2 **a** $\begin{pmatrix} 3 \\ 5 \end{pmatrix}$ **b** $\begin{pmatrix} -2 & 1 \\ -4 & 7 \end{pmatrix}$

3 **a** $\begin{pmatrix} -3 & -2 & -1 \\ 3 & 3 & 0 \end{pmatrix}$ **b** $\begin{pmatrix} 1 & -4 \\ 0 & 9 \end{pmatrix}$

4 **a** Not possible **b** $\begin{pmatrix} -6 & -4 \\ -3 & -2 \end{pmatrix}$
c Not possible **d** $\begin{pmatrix} 7 \\ 0 \end{pmatrix}$
e (-8) **f** $(-7 \ \ -7)$

5 $\begin{pmatrix} 2 & 6-a & 2a \\ 1 & 4 & -2 \end{pmatrix}$

6 $\begin{pmatrix} 3x+2 & 0 \\ 0 & 3x+2 \end{pmatrix}$

7 **a** $\begin{pmatrix} 1 & 4 \\ 0 & 1 \end{pmatrix}$ **b** $\begin{pmatrix} 1 & 6 \\ 0 & 1 \end{pmatrix}$ **c** $\begin{pmatrix} 1 & 2 \times k \\ 0 & 1 \end{pmatrix}$

8 **a** $\begin{pmatrix} a^2 & 0 \\ ab & 0 \end{pmatrix}$ **b** 3

9 **a** $\begin{pmatrix} -8 & -14 \\ -4 & -7 \\ 0 & 0 \end{pmatrix}$ **b** $(-16 \ \ 29)$

10 **a** $\begin{pmatrix} -1 \\ 1 \\ -2 \end{pmatrix}$ **b** $(-3 \ \ 2 \ \ 3)$

Exercise 4D

1 **a** Not linear **b** Not linear **c** Not linear
d Linear **e** Not linear **f** Linear

2 **a** Linear: $\begin{pmatrix} 2 & -1 \\ 3 & 0 \end{pmatrix}$ **b** Not linear
c Not linear **d** Linear: $\begin{pmatrix} 0 & 2 \\ -1 & 0 \end{pmatrix}$
e Linear: $\begin{pmatrix} 0 & 1 \\ 1 & 0 \end{pmatrix}$

3 **a** Not linear **b** Linear: $\begin{pmatrix} 0 & -1 \\ 1 & 0 \end{pmatrix}$
c Linear: $\begin{pmatrix} 1 & -1 \\ 1 & -1 \end{pmatrix}$ **d** Linear: $\begin{pmatrix} 0 & 0 \\ 0 & 0 \end{pmatrix}$
e Linear: $\begin{pmatrix} 1 & 0 \\ 0 & 1 \end{pmatrix}$

4 **a** $\begin{pmatrix} 2 & 1 \\ 0 & -1 \end{pmatrix}$ **b** $\begin{pmatrix} 0 & -1 \\ 1 & 2 \end{pmatrix}$

5 **a** (1, 1), (−2, 3), (−5, 1)
b (3, −2), (14, −6), (9, −2)
c (−2, −2), (−6, 4), (−2, 10)

6 **a** (−2, 0), (0, 3), (2, 0), (0, −3)
b (−1, −1), (−1, 1), (1, 1), (1, −1)
c (−1, −1), (1, −1), (1, 1), (−1, 1)

Exercise 4E

1 **a** Reflection in x-axis (or line $y = 0$)
b Rotation 90° anticlockwise about (0, 0)
c Rotation 90° clockwise about (0, 0)

2 **a** Enlargement scale factor $\frac{1}{2}$, centre (0, 0)
b Reflection in line $y = x$
c No change (this is the Identity matrix)

3 **a** Rotation 45° clockwise about (0, 0)
b Enlargement scale factor 4, centre (0, 0)
c Rotation 225° anticlockwise about (0, 0)

4 **a** $\begin{pmatrix} 0 & 1 \\ -1 & 0 \end{pmatrix}$ **b** $\begin{pmatrix} 1 & 0 \\ 0 & -1 \end{pmatrix}$ **c** $\begin{pmatrix} 2 & 0 \\ 0 & 2 \end{pmatrix}$

5 **a** $\begin{pmatrix} -4 & 0 \\ 0 & -4 \end{pmatrix}$ **b** $\begin{pmatrix} 0 & 1 \\ 1 & 0 \end{pmatrix}$ **c** $\begin{pmatrix} -\frac{1}{\sqrt{2}} & -\frac{1}{\sqrt{2}} \\ \frac{1}{\sqrt{2}} & -\frac{1}{\sqrt{2}} \end{pmatrix}$

Exercise 4F

1 **a** $\begin{pmatrix} 0 & 1 \\ 1 & 0 \end{pmatrix}$; Reflection in $y = x$
b $\begin{pmatrix} 0 & 1 \\ 1 & 0 \end{pmatrix}$; Reflection in $y = x$
c $\begin{pmatrix} -2 & 0 \\ 0 & -2 \end{pmatrix}$; Enlargement scale factor −2, centre (0, 0)
d $\begin{pmatrix} 1 & 0 \\ 0 & 1 \end{pmatrix}$; Identity (no transformation)
e $\begin{pmatrix} 4 & 0 \\ 0 & 4 \end{pmatrix}$; Enlargement scale factor 4, centre (0, 0)

2 **a** $A = \begin{pmatrix} 0 & -1 \\ 1 & 0 \end{pmatrix}$, $B = \begin{pmatrix} -1 & 0 \\ 0 & -1 \end{pmatrix}$
$C = \begin{pmatrix} 1 & 0 \\ 0 & -1 \end{pmatrix}$, $D = \begin{pmatrix} -1 & 0 \\ 0 & 1 \end{pmatrix}$
b **i** Reflection in y-axis
ii Reflection in y-axis
iii Rotation of 180° about (0, 0)
iv Reflection in $y = -x$
v No transformation (Identity)
vi Rotation of 90° anticlockwise about (0, 0)
vii No transformation (Identity)

3 Reflection in $y = x$

5 **a** $\begin{pmatrix} 0 & -1 \\ 1 & 0 \end{pmatrix}$
b Rotation of 90° anticlockwise about (0, 0)
c Rotation of 45° anticlockwise about (0, 0)
d $\begin{pmatrix} 1 & 0 \\ 0 & 1 \end{pmatrix}$ (Identity matrix)

6 **a** $\begin{pmatrix} -1 & 0 \\ 0 & 1 \end{pmatrix}$ **b** Reflection in y-axis

7 $\begin{pmatrix} 0 & -1 \\ -1 & 0 \end{pmatrix}$, Reflection in the line $y = -x$

8 $\begin{pmatrix} 8 & 0 \\ 0 & 8 \end{pmatrix}$, Enlargement scale factor 8

Exercise 4G

1 **a** Non-singular. Inverse = $\begin{pmatrix} 1 & 0.5 \\ 2 & 1.5 \end{pmatrix}$
b Singular **c** Singular
d Non-singular. Inverse = $\begin{pmatrix} -5 & 2 \\ 3 & -1 \end{pmatrix}$
e Singular
f Non-singular. Inverse = $\begin{pmatrix} -0.2 & 0.3 \\ 0.6 & -0.4 \end{pmatrix}$

2 **a** −3 **b** −5 **c** $\frac{1}{4}$

3 **a** $\begin{pmatrix} -(2+a) & 1+a \\ 1+a & -a \end{pmatrix}$
b $\begin{pmatrix} -\frac{1}{a} & -\frac{3}{a} \\ \frac{1}{b} & \frac{2}{b} \end{pmatrix}$ (provided $a \neq 0, b \neq 0$)

4 **b** $\begin{pmatrix} 3 & 4 \\ -1 & -1 \end{pmatrix}$

5 **a** $\mathbf{B} = \mathbf{A}^{-1}\mathbf{C}$ **b** $\begin{pmatrix} 1 & 4 \\ -1 & 2 \end{pmatrix}$

6 **a** $\mathbf{A} = \mathbf{C}^{-1}$ **b** $\begin{pmatrix} 2 & -3 \\ -3 & 5 \end{pmatrix}$

7 $\begin{pmatrix} 2 & 4 & -3 \\ 0 & 1 & 2 \end{pmatrix}$

8 $\begin{pmatrix} 1 & 3 \\ -2 & 1 \\ 0 & -1 \end{pmatrix}$

9 **a** $\frac{1}{2ab}\begin{pmatrix} 2b & -b \\ -4a & 3a \end{pmatrix}$ **b** $\begin{pmatrix} -3 & 2 \\ -1 & \frac{3}{2} \end{pmatrix}$

10 **a** $\det(\mathbf{A}) = 0$, $\det(\mathbf{B}) = 0$ **b** $\begin{pmatrix} 0 & 0 \\ 0 & 0 \end{pmatrix}$

Exercise 4H

1 **a** Rotation of 90° anticlockwise about (0, 0)
b $\begin{pmatrix} 0 & 1 \\ -1 & 0 \end{pmatrix}$
c Rotation of −90° anticlockwise about (0, 0)

2 **a** **i** Rotation of 180° about (0, 0)
iii Rotation of 180° about (0, 0)
b **i** Reflection in $y = -x$
iii Reflection in $y = -x$
c $\det(\mathbf{S}) = 1$, $\det(\mathbf{T}) = -1$

3 **a** $\begin{pmatrix} 1 & 0 \\ 0 & -1 \end{pmatrix}$; reflection in $y = 0$
b $\begin{pmatrix} 1 & 0 \\ 0 & -1 \end{pmatrix}$; reflection in $y = 0$
c $\begin{pmatrix} -1 & 0 \\ 0 & 1 \end{pmatrix}$; reflection in $x = 0$
d $\begin{pmatrix} -1 & 0 \\ 0 & 1 \end{pmatrix}$; reflection in $x = 0$

Exercise 4I

1 **a** (0, 0), (−1, 3), (7, 19), (8, 16)
b 40

2 **a** (1, 2), (6, 2), (3, −1)
c 3.75

3 **a** (2, −1), $(3a - 9, -3a)$, (−8, 4), $(3 - 3a, 3 + 3a)$
b $-a - 3$
c 2

4 **a** 70 **b** 30 **c** 15
d 90 **e** 90 **f** 210

5 **a** $a^2 + 2a - 9$ **b** −5, −3, 1 or 3

Exercise 4J

1 **a** $x = 3, y = -5$ **b** $x = 0.5, y = 3$

2 **a** $x = 2, y = -3$ **b** $x = -1, y = 4$

Mixed exercise 4K

1 $P = (7, -15)$, $Q = (2, -2)$, $R = (-4, 12)$

2 $\begin{pmatrix} 1 & 4 & 3 \\ -1 & 1 & -2 \end{pmatrix}$

3 **a** $\begin{pmatrix} 0 & 1 \\ 1 & 0 \end{pmatrix}$ **b** Reflection in $y = x$
c $\begin{pmatrix} 1 & 0 \\ 0 & 1 \end{pmatrix}$ (Identity matrix)

4 **a** $\begin{pmatrix} 0 & -1 \\ 1 & 0 \end{pmatrix}$ **b** $\begin{pmatrix} -2 & 3 \\ -1 & 2 \end{pmatrix}$
c $\begin{pmatrix} 1 & 0 \\ 0 & 1 \end{pmatrix}$ (Identity matrix)

5 **a** $\begin{pmatrix} 2 & 0 \\ 0 & -2 \end{pmatrix}$; reflection in x-axis and enlargement s.f. 2, centre (0, 0)
b $\begin{pmatrix} \frac{1}{2} & 0 \\ 0 & -\frac{1}{2} \end{pmatrix}$; reflection in x-axis and enlargement s.f. $\frac{1}{2}$, centre (0, 0)

6 **a** $2p^2 - p$ **b** -1 or $\frac{3}{2}$

7 **a** $\begin{pmatrix} \frac{3}{a} & -\frac{1}{a} \\ -\frac{2}{b} & \frac{1}{b} \end{pmatrix}$ **b** $\begin{pmatrix} -1 & 1 \\ 4 & -1 \end{pmatrix}$

8 **a** $\mathbf{X} = \mathbf{BAB}^{-1}$ **b** $\begin{pmatrix} 6 & 2 \\ -4 & -3 \end{pmatrix}$

Exercise 5A

1 **a** $1 + 2 + 3 + 4 + 5 + 6 + 7 + 8 + 9 + 10 = 55$
b $3^2 + 4^2 + 5^2 + 6^2 + 7^2 + 8^2 = 199$
c $1^3 + 2^3 + 3^3 + 4^3 + 5^3 + 6^3 + 7^3 + 8^3 + 9^3 + 10^3 = 3025$
d $5 + 11 + 21 + 35 + 53 + 75 + 101 + 131 + 165 + 203 = 800$
e $1 + 64 + 225 + 484 + 841 + 1296 = 2911$
f $-2 - 52 - 198 - 488 = -740$

2 **a** $6 + 13 + 20 + ... + (7n - 1)$
b $3 + 17 + 55 + ... + (2n^3 + 1)$
c $-15 - 12 - 7 + ... + (n - 4)(n + 4)$
d $18 + 28 + 40 + ... + k(k + 3)$

3 **a** Statement is true **b** Statement is true
c Statement is not true **d** Statement is true
e Statement is not true

4 Various answers

Exercise 5B

1 **a** 666 **b** 4950 **c** 1495
d 15 150 **e** 3240

2 **b** 32

3 **a** $n(2n - 1)$

5 **b** 3276

Exercise 5C

1 **a** 4565 **b** −28 485 **c** 2576

3 **b** 51

4 $a = 7, b = -3$

5 **b** 14 949

Exercise 5D

2 **a** 1, 3, 6 and 10
b 1, 9, 36 and 100
c The results for **b** are the square of the results for **a**

3 **a** 338 350 **b** 19 670
c 216 225 **d** 981 225

4 **a** 48 230 **b** 672 399
c 332 825 **d** $\frac{(k+1)}{6}(k + 2)(2k + 3)$
e $n^2(2n - 1)^2$

5 **a** $\frac{n}{3}(2n + 1)(4n + 1)$ **b** $\frac{(n^2 - 1)n^2(2n^2 - 1)}{6}$
c $\frac{n}{3}(2n - 1)(4n - 1)$ **d** $\frac{(n + 1)^2(n + 2)^2}{4}$
e $n^2(4n + 1)(5n + 2)$

7 **b** 3 159 675

Exercise 5E

1 **a** 9425 **b** 25 420 **c** 10 507 320 **d** 393 825

2 **a** $\frac{n}{6}(n+1)(2n+13)$

b $\frac{n}{2}(n+1)(n^2+n-1)$

c $\frac{n}{3}(n+1)(2n+1)(6n+1)$

3 **a** $(1\times 2)+(2\times 3)+(3\times 4)+\ldots+n(n+1)$

b 75 640

4 **b** 51 660

5 **b** 1 062 000

6 **a** $\frac{n}{4}(n^3+2n^2+n-4)$

b $\frac{n}{3}(4n^2-1)$

c $\frac{n}{12}(n+1)(n+2)(3n+5)$

7 **b** 235 600

8 **b** 16 170

9 **a** $(7\times 23)+(8\times 26)+(9\times 29)+(10\times 32)+(11\times 35)+(12\times 38)=1791$

10 $\dfrac{n(n+1)(4n-1)}{6}$

Mixed exercise 5F

1 **a** $5+13+33+\ldots+(2n+3^n)$

b $n(n+1)+\frac{3}{2}(3^n-1)$

2 **a** 9175 **b** 44 240 **c** 7 843 716

3 **a** n^2+2n-3 **b** $2n+3$ **c** $3(n+1)^2$

4 27 900

5 $\frac{n}{4}(n+1)(n^2-3n-2)$

7 **b** $\frac{n}{6}(n+1)(2n+7)\log 2$

9 **b** 21 049

10 **b** 5 645 178

11 **a** $n^2(n+1)$ **b** $n(n^2+7n+16)$

c $\frac{n^2}{2}(n^2+2n-1)$

12 **b** 65 720

13 **b** 740 430

14 **b** $\frac{n}{6}(n+1)(2n+1)$

15 **c** −33 200

Exercise 6B

9 **a** $7(13^k)$

10 **a** $144k-198$

Mixed exercise 6E

2 **a** $\mathbf{B}^2=\begin{pmatrix}1&0\\0&9\end{pmatrix}$, $\mathbf{B}^3=\begin{pmatrix}1&0\\0&27\end{pmatrix}$

4 **a** 7, 29, 133, 641

5 **b** $\begin{pmatrix}1-8n&-16n\\4n&8n+1\end{pmatrix}$

8 **a** $2, \frac{5}{4}, \frac{11}{16}, \frac{17}{64}, -\frac{13}{256}$

Review exercise 2

1 **a** Does not exist; the number of columns in **A** is not equal to the number of rows in **B**.

b $\begin{pmatrix}6&4&2\\9&4&4\end{pmatrix}$ **c** $\begin{pmatrix}14\\28\end{pmatrix}$

d Does not exist; the number of columns in **C** is not equal to the number of rows in **BA**.

2 $a=-2, b=3$

4 $bc-ad$

5 **a** $-\frac{2}{3}$ **b** −2 **c** −4

6 **a** $\begin{pmatrix}-1&-1\\-3&-2\end{pmatrix}$ **b** $\begin{pmatrix}76&-33\\-99&43\end{pmatrix}$

7 **a** $2k^2+3k-3$ **b** $-\frac{7}{2}$ or 2

8 $\begin{pmatrix}3&0\\7&5\end{pmatrix}$

9 **a** $\begin{pmatrix}1&\frac{1}{2}\\3&2\end{pmatrix}$ **b** $\begin{pmatrix}2&1\\3p+3&2p+\frac{3}{2}\end{pmatrix}$

c $-\frac{1}{2}$

10 **c** $x=-7, y=-17$

11 **a** $\frac{1}{35}\begin{pmatrix}5&2\\-5&5\end{pmatrix}$ **b** $\lambda_1=6, \lambda_2=-1$

12 **a** $\frac{1}{pq}\begin{pmatrix}q&q\\3p&4p\end{pmatrix}$ **b** $\frac{1}{pq}\begin{pmatrix}pq&4q^2\\2p^2&13pq\end{pmatrix}$

13 **a** $\begin{pmatrix}4&6\\3&10\end{pmatrix}$ **b** $\begin{pmatrix}-3&3\\-6&3\end{pmatrix}$ **c** $\begin{pmatrix}-9&0\\0&-9\end{pmatrix}$

d Enlargement scale factor −9, centre (0, 0)

14 **a** $\begin{pmatrix}-\frac{1}{\sqrt{2}}&-\frac{1}{\sqrt{2}}\\-\frac{1}{\sqrt{2}}&\frac{1}{\sqrt{2}}\end{pmatrix}$

15 **a** $a=3, b=-4, c=2, d=-3$

c $p=36, q=25$

16 **a** $\begin{pmatrix}-1&2\\0&3\end{pmatrix}$

b $A(2, 1), B(0, 5), C(-2, 4)$

c

17 **a** $\begin{pmatrix}2&0\\0&-2\end{pmatrix}$

b Reflection in x-axis, followed by enlargement scale factor 2, centre (0, 0)

c (6, 0)

18 **a** $\mathbf{C}=\begin{pmatrix}-1&-7\\2&12\end{pmatrix}$ **b** $\mathbf{D}=\begin{pmatrix}1&2\\4&10\end{pmatrix}$

d $\dfrac{2m}{1+m}$

19 **a** **L** represents rotation through 90°, anticlockwise about (0, 0)
M represents an enlargement scale factor 2, centre (0, 0)

c $\theta=45°, k=\sqrt{2}$ **d** $\begin{pmatrix}16&0\\0&16\end{pmatrix}$

20 **b** $\begin{pmatrix}3&1\\-1&3\end{pmatrix}$

25 **b** 17 730

26 8841

27 46 850

29 **b** 957 700
30 **b** 61 907
31 **b** 32 480
32 **b** 26 660
33 **b** 1 805 040
34 **b** -6
36 **b** $p = 13, q = 7$
37 **a** $p = 3, q = -1, r = -2$ **b** 23 703 950
38 **b** 247.5
52 **a** $24 \times 2^{4(n+1)} + 3^{4(n+1)} - 24 \times 2^{4n} - 3^{4n}$
56 **a** $p = 6, q = -8$

Examination style paper

1 $a = 1, b = 4, c = 5$
2 **a** 1.29
3 **a** **R** represents a rotation of 135° anti-clockwise about (0, 0)
S represents an enlargement scale factor $\sqrt{2}$, centre (0, 0)
b $\begin{pmatrix} -1 & -1 \\ 1 & -1 \end{pmatrix}$ **c** 2
4 **a** $1 - \text{i}$ **b** $z^2 - 2z + 2$
c $-1 + 2\text{i}$ and $-1 - 2\text{i}$
6 **a** $\dfrac{1}{x^2 + 6x - 7}\begin{pmatrix} x + 4 & 7 - 2x \\ 1 & x \end{pmatrix}$ **b** -7 or 1
7 **a** $z = 4 + 3\text{i}, w = -3 + 4\text{i}$
b 5 **c** 2.21
d

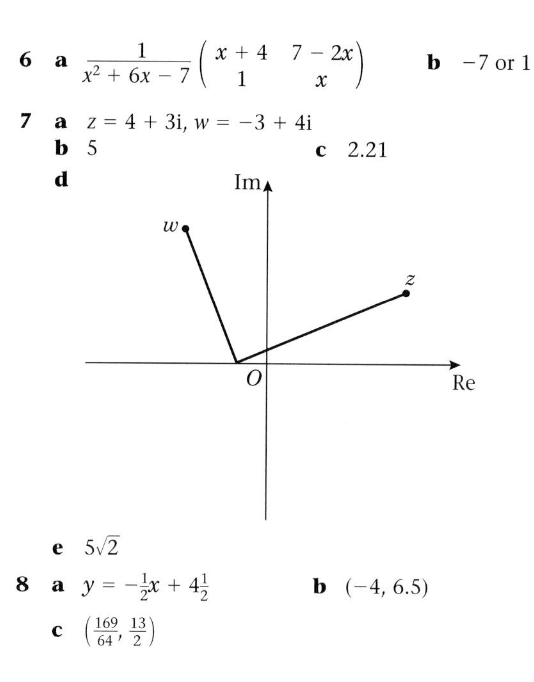

e $5\sqrt{2}$
8 **a** $y = -\frac{1}{2}x + 4\frac{1}{2}$ **b** $(-4, 6.5)$
c $\left(\frac{169}{64}, \frac{13}{2}\right)$

Index